INTERNATIONAL SYMPOSIUM ON COGNITIVE BIOLOGY

February 6 – 8, 2026

COGNITION ACROSS LIFE
FROM CELL TO CONSCIOUSNESS

Organizers:
Kuruom School of Advanced Sciences (KSAS), India
&
Institute of Advanced Sciences (INADS), USA

Venue:
Kuruom School of Advanced Sciences
Gata No. 960, Mohanlalganj, Lucknow, Uttar Pradesh 226301, India

Table of Contents

Program

Day 1: February 6, 2026

4:00 pm
Social

5:30 pm – 6:30 pm
Performance by the students of Kuruom Vidyalaya on cognition and life
Welcome Dance
Guitar played by Dr. John Bogdan

- Play 1: Voice of Mind
- Play 2: Operation Sindoor

6:30 pm – 7:30 pm
Symposium Speech
Chair: Dr. Alex Hankey, Theoretical Physicist, Professor of Philosophy of Science, MIT World Peace University, Pune, India
Topic: *Framing the Vision of Cognitive Biology*
Dr. Suhas Kshirsagar, Advisor, National Ayurvedic Medical Association

7:30 pm
Dinner

Day 2: February 7, 2026

9:00 am – 10:00 am
Registration, Social, and Breakfast

10:00 am – 10:45 am
Saraswati Vandana followed by a formal welcome and presentation by Dr. Raj Kumar, Kuruom School of Advanced Sciences (KSAS), India

Session I

10:45 am – 11:00 am
Speech by Vishist Athiti Mr. Raja Ram

11:00 am – 12:00 pm

Keynote Speech

Chair: Dr. John Bogdan, President and CEO, Valaria Technical Consultants, LLC, MD, USA

Topic: *Cognition Beyond Brains: Redefining Intelligence in Nature*

Dr. Shiv Satchit, Journalist and Media Technology Consultant, UK

12:00 pm – 12:15 pm

Break

Session II

Minimal Cognition: Insight from Unicellular Organisms and Plants

Chair: Dr. Ashutosh Pareek, Educator and Author, India

12:15 pm – 12:45 pm

Presentation I: Dr. Yamini Bhushan Tripathi, Retired Professor, Banaras Hindu University (BHU), India

Topic: *Cognition of Medhya Rasayana*

12:45 pm – 1:15 pm

Presentation II: Mr. Harsh Pandey, Student, Jawaharlal Nehru University (JNU), India

Topic: *Achar Rasayan: The Psychosomatic Link Between Behavioral Ethics and Biological Longevity*

1:15 pm – 1:45 pm

Presentation III: Dr. Bal Ram Singh, President, Institute of Advanced Sciences (INADS), MA, USA

Topic: *Role of Atomic, Molecular, and Genetic Factors in Evolutionary Conscious Connection of Microorganisms and Plants*

1:45 pm – 2:45 pm

Lunch Break

Session III

The Origins of Cognition: Evolutionary and Developmental Perspective

Chair: Dr. Yamini Bhushan Tripathi, Retired Professor, Banaras Hindu University (BHU), India

2:45 pm – 3:15 pm

Presentation I: Dr. Chander Kanwar Bhardwaj, Life Member, World Association of Vedic Studies, Atlanta, GA, USA

Topic: *Cosmic Cognitive Restructuring to Go Beyond Cognition*

3:15 pm – 3:45 pm

Presentation II: Dr. Ashutosh Pareek, Educator and Author, India

Topic: *Ārṣa Theories of Cognition: Scriptural Foundations and Developmental Perspectives*

3:45 pm – 4:15 pm

Presentation III: Dr. Jerry Solfvin, Professor, University of Massachusetts - Dartmouth, MA, USA

Topic: *Physical Research, Parapsychology, and Cognitive Biology*

4:15 pm – 4:45 pm

Break

Session IV

Panel Discussion

4:45 pm – 5:30 pm

Topic: *Where Does Cognition Begin? Philosophical and Biological Views*

Chair: Dr. Alex Hankey, Theoretical Physicist, Professor of Philosophy of Science, MIT World Peace University, Pune, India

Panelists: Dr. Shiv Satchit, Dr. Chander Kanwar Bhardwaj, Dr. Mrityunjay Guha Majumdar, and Dr. Yamini Bhushan Tripathi

5:30 pm - 6:00 pm

Poster Session

6:30 pm

Dinner

Day 3: February 8, 2026

8:00 am – 9:00 am

Breakfast and Social

Session V

9:00 am – 10:00 am

Keynote Speech

Chair: Dr. Bal Ram Singh, President, INADS, USA

Topic: *Delaying Cognitive Decline: Disease-Modifying Therapy and the Evolving Role of Rehabilitation*

Dr. Nirmal Surya, Chairman, Surya Neuro Center and Fellow of the American Academy of Neurology

10:00 am – 10:15 am

Break

Session VI

Cellular and Molecular Bases of Cognition

Chair: Dr. Gajendra Pratap Singh, Professor, Jawaharlal Nehru University (JNU), Delhi, India

10:15 am – 10:45 am

Presentation I: Dr. Chandra Sekhar Mayanil, Retired Principal Scientist, Lurie Children's Hospital, Chicago, IL, USA

Topic: *Cellular Cognition in Neuro-Immune Systems*

10:45 am – 11:15 am

Presentation II: Dr. Alex Hankey, Theoretical Physicist, Professor of Philosophy of Science, MIT World Peace University, Pune, India

Topic: *Information Encoded by the Subjective Self During Cognition: Its Overall Structure as a Banach Space*

11:15 am – 11:45 am

Presentation III: Dr. John Bogdan, President and CEO, Valaria Technical Consultants, LLC, MD, USA

Topic: *Cellular and Molecular Toxins: Friends or Foes?*

11:45 am – 12:15 pm

Presentation IV: Dr. Abha Sharma, Professor, NIPER, Raebareli, India
Topic: *Molecular Target-Based Development of Drug-Like and Fluorescent Molecules for Alzheimer's Disease*

12:15 pm – 1:15 pm

Lunch

Session – VII

Synthetic and Artificial Systems: Bio-Inspired Cognitive Architecture
Chair: Dr. Chander Kanwar Bhardwaj, Life Member, World Association of Vedic Studies, Atlanta, GA, USA

1:15 pm – 1:45 pm

Presentation I: Dr. Gajendra Pratap Singh, Professor, Jawaharlal Nehru University (JNU), Delhi, India
Topic: *Mathematical Network Models of Cognition in Biological Systems*

1:45 pm – 2:15 pm

Presentation II: Dr. Mrityunjoy Guha Majumdar, Assistant Professor, Amrita Vishwaidyapeetham, Delhi, India
Topic: *Quantum General Artificial Intelligence*

2:15 pm – 2:45 pm

Presentation III: Dr. Vishwanathan Ramesh, Professor, University of Frankfurt, Germany
Topic: *A Transdisciplinary Systems Perspective on Intelligence*

2:45 pm – 3:15 pm

Presentation IV: Dr. Valeri Barsegov, Professor, University of Massachusetts - Lowell, MA, USA
Topic: *Computational Modeling of Biological Systems: Application to Cognition at the Cellular and Whole System Level.*

3:15 pm – 3:30 pm

Break

Session VII

Environmental Influences on Cognitive Processes
Chair: Dr. Shiv Satchit, Journalist and Media Technology Consultant

3:30 pm – 4:00 pm

Presentation I: Dr. K. Gopinath Pillai, Director, Indic Dharma Academy, India
Topic: *Cosmic Intelligence and the Evolution of Cognition*

4:00 pm – 4:30 pm

Presentation II: Dr. Darshan Mehta, Assistant Professor, Harvard Medical School, Boston, MA, USA
Topic: *Mitochondria Under Pressure: Exploring the Stress-Mind-Body Connection in Health and Disease*

4:30 pm – 5:00 pm

Presentation III: Dr. Peter Wayne, Assistant Professor, Osher Center, Harvard Medical School, Boston, MA, USA
Topic: *Minding the Body: Tai Chi for Cognitive-Motor Health in Older Adults*

5:00 pm – 5:30 pm

Break

Session – VIII

Panel Discussion

5:30 pm – 6:15 pm

Topic: *Designing a Research Agenda for Cognitive Biology*
Chair: Dr. Bal Ram Singh, President, INADS, USA
Panelists: Dr. John Bogdan, Dr. Ashutosh Pareek, and Dr. Vishwanathan Ramesh

6:15 pm – 6:30 pm

Vote of Thanks and Conclusion

Cognition Across Life: From Cells to Consciousness

Dr. Raj Kumar

Kuruom School of Advanced Sciences, Mohanlalganj, Lucknow, India

Cognition is traditionally associated with brains, neurons, and nervous systems. In everyday language and in much of scientific discourse, cognition refers to the processes by which organisms perceive their environment, think, make decisions, learn from experience, and adapt their behaviour. This conventional definition places cognition firmly within the domains of neuroscience and psychology and implicitly assumes that cognition emerged only after the evolution of specialized neural tissues. Within this framework, the brain is regarded as the organ of cognition, and neurons are considered its fundamental functional units. This perspective has been extraordinarily productive, enabling major advances in understanding perception, memory, learning, behaviour, and consciousness. It has revealed how neurons communicate through synapses, how neural circuits encode information, and how distributed neural activity gives rise to perception and decision-making. However, this definition of cognition is fundamentally anatomical rather than functional. When cognition is defined anatomically, it becomes restricted to organisms possessing nervous systems. When defined functionally—as the ability of a system to acquire information, integrate that information, store it, and generate adaptive responses—it becomes clear that cognition is not restricted to brains but exists across all forms of life. From this broader biological perspective, cognition is not a late evolutionary innovation but a fundamental property of living systems themselves (Lyon, 2006; Mitchell et al., 2009; Levin, 2019; Baluška and Levin, 2016).

This shift from an anatomical to a functional definition of cognition reflects advances across multiple scientific disciplines, including microbiology, cell biology, systems biology, developmental biology, and biophysics. These fields have demonstrated that living cells continuously monitor their internal and external environments, integrate multiple streams of information, and generate adaptive responses that promote survival, stability, and function. Cells sense chemical gradients, mechanical forces, temperature changes, electrical signals, nutrient availability, and molecular indicators of stress or damage. They integrate these diverse signals through complex intracellular signaling networks and gene regulatory systems. Based on this integrated information, cells select among alternative responses such as migration, proliferation, differentiation, repair, secretion, or programmed cell death. These responses are not rigid or predetermined but depend on both current environmental conditions and the cell's prior history. Thus, cognition emerges naturally from the fundamental regulatory architecture of living systems, in which information processing and adaptive decision-making are intrinsic properties of cellular function (Bray, 2009; Alon, 2007; Lim et al., 2014).

At its core, cognition involves information processing. Living systems exist far from thermodynamic equilibrium and must continuously process information to maintain their structural and functional integrity. Cells accomplish this through molecular sensing mechanisms embedded in their membranes and cytoplasm. Receptors on the cell surface detect extracellular signals such as growth factors, cytokines, hormones, neurotransmitters, nutrients, and even toxins. Intracellular sensors monitor metabolic state, oxidative stress, DNA integrity, protein folding, and energy availability. Detection of these signals initiates intracellular signaling cascades involving kinases, phosphatases, ion channels, second messengers such as calcium and cyclic AMP, and transcription factors. These signaling cascades form interconnected networks capable of integrating multiple simultaneous inputs. The outcome of signaling depends not on individual signals alone but on the overall network state, allowing cells to interpret information contextually rather than react reflexively. This integration of information is a fundamental hallmark of cognitive systems (Bray, 2009; Lim et al., 2014) (Fig. 1).

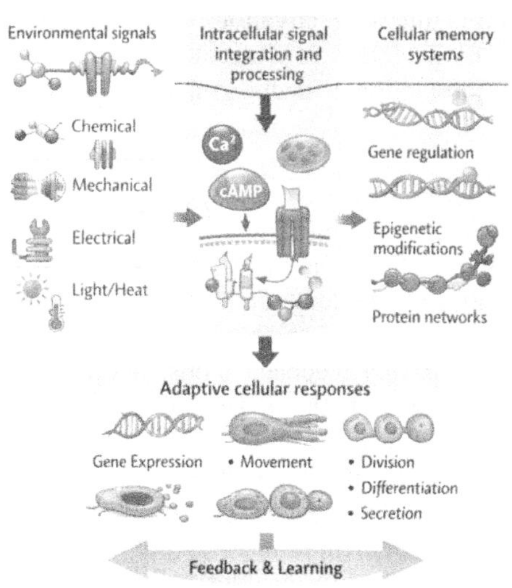

Figure 1: Cellular cognition as an information-processing system.

Cellular signaling networks exhibit computational properties analogous to electronic circuits. These networks implement logical operations that enable cells to evaluate complex combinations of signals and generate appropriate responses. Logical AND operations require multiple signals to be present simultaneously for activation. For example, T lymphocyte activation requires both antigen recognition and costimulatory signaling, ensuring that immune responses occur only under appropriate conditions. Activation of T lymphocytes requires simultaneous recognition of antigen through the T cell receptor and engagement of costimulatory receptors such as CD28. Neither signal alone is sufficient to activate the immune response, preventing inappropriate immune activation and ensuring specificity (Smith-Garvin et al., 2009). Logical OR operations allow multiple independent signals to trigger similar responses, enabling flexibility in responding to diverse environmental conditions. Stress response pathways can be activated by multiple independent stimuli, including oxidative stress, heat shock, or DNA damage, allowing flexible adaptive responses to diverse threats (Alon, 2007). Logical NOT operations allow inhibitory signals to suppress responses, preventing inappropriate activation. For example, the tumor suppressor protein p53 promotes cell cycle arrest or apoptosis in response to stress, but its activity is inhibited by regulatory proteins such as MDM2 under normal conditions, preventing unnecessary activation of cell death pathways (Levine, 1997). Feedback loops within signaling networks enable amplification, stabilization, and adaptation. Positive feedback can stabilize specific cellular states, while negative feedback enables dynamic regulation and prevents runaway activation. These regulatory architectures allow cells to perform sophisticated information processing analogous to computational decision-making systems (Alon, 2007; Bray, 2009).

Memory is another defining feature of cognition, and cells possess multiple mechanisms for storing information about past experiences. Cellular memory exists in biochemical, genetic, epigenetic, and structural forms. Gene regulatory networks create stable patterns of gene expression that define cellular identity. These stable patterns are maintained through feedback loops that preserve transcriptional states across time. Epigenetic modifications, including DNA methylation and histone modification, allow cells to record past environmental exposures and maintain altered gene expression patterns even after the original signals disappear. These mechanisms allow cells to "remember" developmental history, environmental stress, and signaling events. Such memory enables cells to respond differently to the same stimulus depending on prior experience, a fundamental characteristic of cognitive systems (Bird, 2007; Davidson and Erwin, 2006).

Cells also exhibit learning-like behavior through adaptive regulatory changes. For example, bacterial chemotaxis systems adjust sensitivity based on prior exposure to chemical gradients, allowing cells to respond effectively across a wide range of signal intensities. This process, known as adaptation, enables cells to detect changes relative to background conditions rather than absolute signal levels (Wadhams and Armitage, 2004). Similarly, repeated stimulation can alter cellular signaling responses through receptor desensitization, gene expression changes, or structural remodeling. These adaptive changes allow cells to modify behavior based on experience, representing a fundamental form of learning at the cellular level (Mitchell et al., 2009).

Importantly, cellular cognition is not limited to individual cells acting in isolation but also emerges from collective interactions among cellular populations. Life depends fundamentally on cooperation among cells. In multicellular organisms, trillions of cells coordinate their activities through biochemical signaling, mechanical interactions, and bioelectrical communication. Each cell operates as an autonomous information-processing unit while simultaneously participating in distributed regulatory networks that govern tissue and organismal function. This collective cognition allows cells to coordinate growth, differentiation, repair, and physiological regulation. Cooperation ensures that cellular behavior contributes to the stability and survival of the organism as a whole. Without such cooperative information processing, multicellular life would not be possible (Levin, 2014; Baluška and Levin, 2016).

Collective cognition is evident even in microbial populations. Bacteria communicate through quorum sensing, a process in which cells release signaling molecules and detect their concentration in the environment. When signal concentration reaches a threshold indicating sufficient population density, coordinated gene expression occurs across the population. This enables collective behaviors such as biofilm formation, virulence activation, and metabolic coordination. These behaviors emerge not from centralized control but from distributed information processing across the population, demonstrating collective decision-making at the cellular level (Bassler and Losick, 2006).

Collective cognition also plays a central role in development and regeneration. During embryonic development, cells coordinate their behavior to generate complex anatomical structures with precise spatial organization. This process requires continuous communication and information exchange among cells. Bioelectrical signaling networks, mediated by ion channels and gap junctions, regulate pattern formation and tissue organization. These networks enable cells to coordinate behavior across large spatial scales, ensuring that tissues develop correct structure and function. Remarkably, these regulatory systems can restore anatomical structure after injury, demonstrating goal-directed regulation at the tissue level (Levin, 2014; Levin, 2019).

Even within the nervous system, cognition is not exclusively a neuronal phenomenon. The brain contains large populations of non-neuronal cells, including astrocytes, oligodendrocytes, and microglia. These glial cells actively participate in information processing, synaptic regulation, and neural function. Astrocytes detect neurotransmitter release, regulate synaptic transmission, and coordinate neural network activity through calcium signaling and gliotransmitter release. Microglia monitor neural tissue, detect injury or infection, and regulate inflammation and synaptic remodeling. Oligodendrocytes regulate myelination, influencing signal transmission speed and neural plasticity. These cells exhibit sensing, integration, and adaptive response—core features of cognition—demonstrating that cognition within the brain is distributed across multiple interacting cell types rather than restricted to neurons alone (Fields et al., 2015; Allen and Lyons, 2018; Barres, 2008).

Single-celled organisms provide particularly striking examples of cellular cognition. The slime mold *Physarum polycephalum*, despite lacking a nervous system, can navigate complex environments and solve spatial optimization problems. It forms efficient networks connecting food sources, demonstrating distributed problem-solving ability. It can also store information about environmental conditions and use that information to guide future behavior (Nakagaki et al., 2000). Similarly, ciliates such as *Stentor roeselii* exhibit adaptive behavioral responses to repeated stimulation. They modify their responses based on prior experience, demonstrating learning-like behavior at the single-cell level (Jennings, 1906; Dexter et al., 2019).

From an evolutionary perspective, cognition did not originate with neurons but emerged from fundamental cellular information-processing systems that existed long before nervous systems evolved. Neurons represent specialized adaptations that enhance communication speed, coordination, and computational capacity. The basic mechanisms of signal detection, information integration, memory storage, and adaptive response existed in cellular regulatory networks before the evolution of nervous systems. Nervous systems represent an evolutionary scaling of cellular cognition rather than its origin (Jékely, 2021; Levin, 2019).

Despite the growing body of evidence supporting the view that cognition is a fundamental property of living systems, it is essential to clarify the conceptual boundaries and avoid potential misunderstandings associated with this expanded definition. The attribution of cognitive properties to cells does not imply that cells possess subjective experience, awareness, or consciousness in the same sense as organisms with nervous systems. Rather, cellular cognition refers specifically to functional capabilities such as information acquisition, signal integration, memory storage, and adaptive response. These capabilities can be rigorously defined in terms of measurable biochemical, biophysical, and regulatory processes without invoking subjective phenomenology. It is therefore important to distinguish between functional cognition, which can be objectively studied across biological systems, and conscious cognition, which involves subjective awareness and remains a distinct and more complex phenomenon (Lyon, 2006; Levin, 2019; Jékely et al., 2021).

A second caveat concerns the risk of anthropomorphism. Describing cellular behavior using terms such as "decision-making," "memory," or "learning" may create the impression that cells possess intentionality or agency in the psychological sense. However, these terms are used in a precise scientific context to describe regulatory dynamics that are formally equivalent to information processing and adaptive regulation. Similar terminology is widely used in other scientific fields, such as computer science, control theory, and systems biology, where systems lacking consciousness nonetheless exhibit information processing, state-dependent behavior, and adaptive regulation. The use of cognitive terminology in cellular biology reflects functional similarities in information processing architectures rather than implying equivalence in subjective experience (Mitchell et al., 2009; Bray, 2009; Baluška and Levin, 2016).

It is also important to recognize that cellular cognition operates within physical and biochemical constraints. Cellular information processing is mediated by molecular interactions, regulatory networks, and biophysical mechanisms that are subject to thermodynamic limits, stochastic fluctuations, and evolutionary constraints. Cellular behavior emerges from the dynamics of these regulatory systems rather than from centralized control or abstract computation independent of physical processes. Understanding cellular cognition therefore requires integrating molecular biology, systems biology, and biophysics to explain how information processing emerges from the interaction of molecular networks (Alon, 2007; Lim et al., 2014).

Another important limitation concerns differences in scale, complexity, and integration across biological systems. While cellular information processing provides the foundation for cognition, higher-level cognitive phenomena such as perception, reasoning, and consciousness involve additional layers of organization, including specialized neural circuits, large-scale network integration, and dynamic interactions across multiple spatial and temporal scales. Nervous systems represent evolutionary innovations that dramatically increase the speed, efficiency, and integration of information processing. Thus, while cellular cognition provides the foundational substrate, neural cognition represents a distinct level of biological organization with emergent properties that cannot be fully reduced to individual cellular processes alone (Fields et al., 2015; Levin, 2019; Jékely, 2021).

Finally, the concept of cellular cognition should be viewed as a scientific framework rather than a metaphysical claim. Its value lies in providing a unifying conceptual model for understanding biological regulation, development, adaptation, and disease in terms of information processing and distributed control. This framework enables integration across disciplines, including cell biology, neuroscience, developmental biology, and medicine, while remaining grounded in experimentally observable mechanisms. By recognizing both the explanatory power and the conceptual limits of cellular cognition, researchers can avoid overinterpretation while fully leveraging its potential to advance understanding of biological intelligence across scales (Mitchell et al., 2009; Levin, 2019; Baluška and Levin, 2016).

Understanding cognition as a continuum across biological scales has profound implications for medicine. Diseases such as cancer can be understood not only as genetic disorders but as disruptions of cooperative cellular information processing. Cancer cells lose integration with multicellular regulatory networks and behave more autonomously, prioritizing individual survival over collective function. This perspective suggests therapeutic strategies focused on restoring proper regulatory communication and cellular cooperation rather than simply destroying cancer cells (Levin, 2021).

This perspective also transforms approaches to regenerative medicine. Tissue repair and regeneration depend on coordinated cellular behavior guided by distributed information processing. Understanding how cells interpret signals and coordinate behavior enables the development of therapies that guide regeneration and restore tissue structure and function. Manipulating bioelectrical signaling, gene regulatory networks, and cellular communication pathways provides new approaches to regenerative medicine (Levin, 2019).

In synthetic biology and bioengineering, understanding cellular cognition enables the design of biological systems with predictable behavior. Synthetic morphology seeks to guide cellular organization to engineer tissues, organs, and biological structures. By manipulating cellular signaling networks and information processing systems, researchers can design biological systems capable of performing specific functions (Kamm and Bashir, 2014).

Understanding cellular cognition also bridges the conceptual gap between molecular biology and neuroscience. Neural cognition emerges from cellular information processing systems that evolved long before neurons. The mechanisms underlying perception, learning, and decision-making in neural systems represent specialized extensions of fundamental cellular regulatory processes. Consciousness itself may represent the highest level of integration in hierarchical biological information-processing systems built upon cellular cognition (Levin, 2019).

In summary, cognition is not restricted to brains but is a fundamental property of living systems (Fig. 2). Cells sense their environment, integrate information, store memory, and generate adaptive responses. These processes exist across all levels of biological organization, from single cells to complex organisms. Nervous systems represent specialized adaptations that enhance cognitive capacity but do not represent the origin of cognition. Recognizing cognition as a continuum across biological scales provides a unified framework for understanding intelligence, behavior, development, and disease. This perspective transforms our understanding of life itself, revealing that cognition is not an emergent property of brains alone but a fundamental characteristic of living matter.

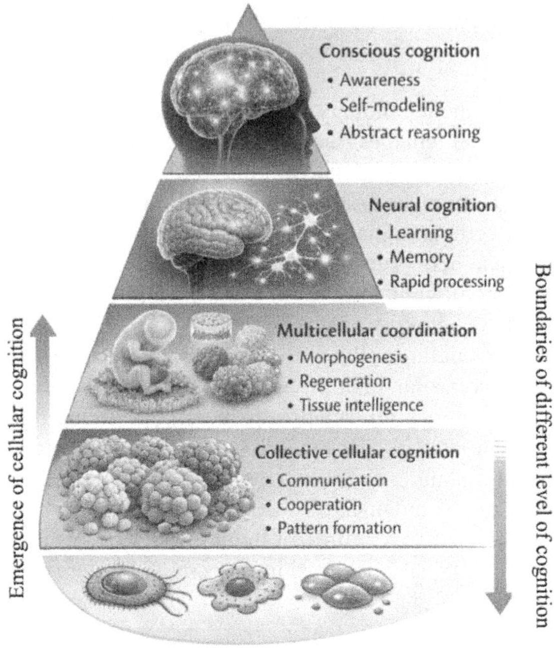

Figure 2: Emergence of cognition from cellular network to consciousness.

Finally, after consideration of cellular intelligence and its caveats, all I can say is "*Competence without comprehension, but with cooperation and networking.*"

References

Allen, N.J., and Lyons, D.A. (2018). Glia as architects of central nervous system formation and function. *Science*, 362, 181–185.

Alon, U. (2007). *An Introduction to Systems Biology: Design Principles of Biological Circuits.* Chapman & Hall/CRC.

Baluška, F., and Levin, M. (2016). On having no head: cognition throughout biological systems. *Frontiers in Psychology*, 7, 902.

Barres, B.A. (2008). The mystery and magic of glia. *Neuron*, 60, 430–440.

Bassler, B.L., and Losick, R. (2006). Bacterially speaking. *Cell*, 125, 237-246.

Bird, A. (2007). Perceptions of epigenetics. *Nature*, 447, 396–398.

Bray, D. (2009). *Wetware: A Computer in Every Living Cell.* Yale University Press.

Davidson, E.H., and Erwin, D.H. (2006). Gene regulatory networks and evolution. *Science*, 311, 796–800.

Dexter, J.P., et al. (2019). Individual cells can learn. *Current Biology*, 29, R1215–R1217.

Fields, R.D., Woo, D.H., and Basser, P.J. (2015). Glial regulation of neural signaling. *Science*, 349, 125–129.

Jennings, H.S. (1906). *Behavior of the Lower Organisms.* Columbia University Press.

Jékely, G. (2021). The chemical brain hypothesis. *Philosophical Transactions of the Royal Society B*, *376*, 20190761.

Kamm, R.D., and Bashir, R. (2014). Creating living cellular machines. *Annals of Biomedical Engineering*, 42, 445–459.

Kandel, E. R., Dudai, Y., & Mayford, M. R. (2014). The molecular and systems biology of memory. *Cell*, 157, 163–186.

Keijzer, F. A. (2015). Moving and sensing without input and output: Early nervous systems and the origins of the animal sensorimotor organization. *Biology & Philosophy*, 30, 311–331.

Levin, M. (2014). Molecular bioelectricity in developmental biology. *Developmental Biology*, 387, 1–15.

Levin, M. (2019). The computational boundary of a "self": Developmental bioelectricity drives multicellularity and scale-free cognition. *Frontiers in Psychology*, 10, 2688.

Levin, M. (2021). Bioelectric signaling in regeneration. *Nature Reviews Molecular Cell Biology*, 22, 195-213.

Lim, W.A., Mayer, B.J., and Pawson, T. (2014). *Cell Signaling: Principles and Mechanisms.* Garland Science.

Lyon, P. (2006). The biogenic approach to cognition. *Cognitive Processing*, 7, 11–29.

Mitchell, A., et al. (2009). Adaptive prediction in microbes. *Nature*, 460, 220–224.

Nakagaki, T., Yamada, H., and Tóth, Á. (2000). Maze-solving by amoeboid organisms. *Nature*, 407, 470.

Nicholson, D. J. (2019). Is the cell really a machine? *Journal of Theoretical Biology*, 477, 108–126.

Smith-Garvin, J.E., Koretzky, G.A., and Jordan, M.S. (2009). T cell activation. *Annual Review of Immunology*, 27, 591–619.

Wadhams, G.H., and Armitage, J.P. (2004). Making sense of chemotaxis. *Nature Reviews Molecular Cell Biology*, 5, 1024–1037.

Framing The Vision of Cognitive Biology

Dr. Suhas Kshirsagar

BAMS, MD (Ayurveda) Advisor, National Ayurvedic Medical Association, India

Abstract

In human existence, sleep stands as a profound mystery-a universal yet deeply personal phenomenon that bridges the worlds of the seen and unseen. "Awakened Sleep" is a state that transcends the basic function of rest, inviting us into a realm of expanded awareness where healing, intuition, and transformation become possible. It is the sleep of *sages*-a state of rest imbued with wakeful presence and spiritual insight. Spiritually, it offers a rare opportunity to dissolve into the field of infinite possibilities, accessing the unbounded source of creativity and wisdom.

Yoga and Vedanta both affirm that deep dreamless sleep is considered the closest we come to our source in pure awareness. Therefore, if you could experience deep sleep instead of blanking out, you would grasp your true self. According to various Vedic philosophies, you would at the same time merge with your source and be free of the illusion, or Maya, of the physical world. Advanced yogis, or even meditators of long experience, attest that they are self-aware in deep sleep. Awareness during deep sleep shows that consciousness is never lost and can be experienced in its pure state, without the participation of the active mind.

Sleep, Mind & Consciousness

Sleep is a subjective experience that we've tried hard to quantify and objectify. While continuing the search for objective information, it's important to consider and explore the qualitative aspects of sleep. After all, sleep is an experience, not a measurement! Every night offers a brand-new experience, based on the impact of everything you've encountered throughout the day. What your mind has absorbed and become in your waking hours is what you'll experience during sleep. You can utilize the experience of awakened sleep to process and transcend those experiences, which leads to better health and well-being. Sleep is viewed as the elixir of youth, but even more than that, it can be the gateway to awakening our spiritual intelligence.

There's a growing body of scientific studies that suggest human consciousness is nonlocal; that is, it's not limited to our brains, bodies, and the ways we think of space and time. Ayurveda and Vedic science have always known that consciousness and deep spiritual experience bring us into unity with all that exists. As we tap into the secrets of consciousness through sleep, we have the opportunity to access health, well-being, and creativity beyond our wildest dreams. Through the capacity to watch ourselves as we dream and sleep, we can enter the witness/observer role that helps us move beyond our limited ideas of who we are. We can come to identify not with our everyday roles and personalities, nor our physical body or thoughts and emotions, but as the ocean of bliss and pure potential that is divine consciousness.

The practices of *Awakened Sleep* center around six main principles, culled from Ayurvedic and Vedic wisdom:

- **Stress Management and Meditation**: We can't avoid stress, but we can learn to manage it more effectively. We emphasize meditation and mindfulness as a crucial way to mitigate our stressors. Awakened sleep is an extension of meditation that heals the effects of stress.
- **Emotional Regulation**: Science is becoming more aware of the intimate connection between mind and body, and of emotional states as a key driver of our physiology. In Ayurveda, it's believed that the majority of physical ailments have an emotional root cause. Understanding and managing our emotions, rather than allowing our emotions to control us, is essential to our health and our sleep.
- **Nutrition**: When the body is balanced, the mind feels better. Ayurveda notes that foods can directly affect our emotional state, and science is discovering that certain nutrients are necessary for generating neurotransmitters that can impact our moods and induce sleep. We'll discuss food and nutrition as it pertains to a healthy sleep regimen.
- **Movement**: In Ayurveda, it's said that movement is life. The ancient Ayurvedic physician Sushruta was the first documented physician to have prescribed regular movement for health. In this book, we discuss how movement not only assists our health during the day but also can be used in the service of awakened sleep.
- **Self-Care**: Self-care is especially important in setting up a daily routine, known as *dinacharya* in Ayurveda.

Ayurveda recognizes the importance of not only *what* we do throughout the day but also *when* we do it. Sleep researchers are also circadian-rhythm researchers, documenting the importance of aligning our internal rhythms with nature's cycles for proper sleep. Ayurvedic self-care practices align with our natural rhythms.

- **Sleep**: Ayurveda has always recognized sleep as a biological necessity that's foundational to life.

Cellular and Molecular Toxins: Friends or Foes?

John Bogdan Jr., PhD

President and CEO, Valaria Technical Consultants, LLC, Maryland, USA

Abstract

Cognitive biology principles rest on the belief that every organism, whether single-celled or multicellular, is continually engaged in systematic acts of sensing and living alongside others in a changing environment, coupled with its own intentional behaviors to survive. Bacterial toxins, conotoxins, reptilian and insect venoms, and plant poisons have evolved to provide a protective barrier between these organisms and the surrounding ecosystem that supports their survival. Their actions have evolved by sensing and selection in response to recognition of external stimuli. Off-target effects that are detrimental to some organisms are unintended but can target (s). These agents exist in nature around the world, and exposure to biological toxins can be deadly to humans as well as other mammalian species and other organisms. Mammals, birds, fish, and insects can be susceptible to these toxins by contact with the skin and/or through exposure through the aerosol, oral, and parenteral routes.

Before the 20th century, significant morbidity and mortality could be attributed to exposure to diphtheria, tetanus, pertussis, botulinum, and anthrax toxins from infection or environmental exposure, and other bacterial toxins, marine and plant toxins like ricin found in the castor bean or poisonous ivy(ies). Vaccines and therapeutics have been developed to protect or treat against several of these agents, especially for infants whose immature immune system makes them more vulnerable to the effects of infection and who are more likely to die from low-level infection and exposure than adults. Detoxified toxins or venom could be the basis for vaccines for bacterial toxins like diphtheria, tetanus, and pertussis, and various snake and insect venoms and plant toxins, and the basis for generating hyper-immune immunoglobulin to treat symptoms during exposure. Toxoids of tetanus and diphtheria toxins have been used as carrier molecules for pediatric and, in some cases, adult polysaccharide-based vaccines to protect against *Haemophilus influenzae type b*, *Streptococcal pneumoniae,* and *Neisseria meningitidis*. Licensed botulinum neurotoxin like BOTOX can be administered in low concentrations to reduce wrinkles, alleviate migraines, and treat cervical dystonia, while much higher amounts are lethal.

This presentation will provide insights on how these deadly agents affect human consciousness and our awareness by coexistence with them, while recognizing how to use these agents themselves or their detoxified forms to restore, sustain, or protect life on our planet. Recognizing why these agents exist and how they act can be a matter of life or death.

Keywords: Toxins, Venoms, Evolution, Botulinum Neurotoxin, Cognition

Introduction

The study of toxins, venoms, and poisons across the biological spectrum sheds light on their significant evolutionary roles in survival. Whether from bacteria, plants, marine organisms, or animals, these agents have evolved to protect organisms from environmental pressures. While these toxins can be life-threatening, their understanding has provided a wealth of knowledge that has directly informed therapeutic strategies, such as vaccines, antidotes, and the development of novel medicines. This paper will explore the evolutionary significance of these agents, how they impact human health, the therapeutic strategies that have been developed to protect or treat exposure, and how they have been used to develop highly impactful vaccines.

Cognitive biology refers to the ability of organisms, from single-celled bacteria to complex mammals, to engage in systematic sensing of their environment (Calow, 1998). Every organism, whether unicellular or multicellular, has evolved specific mechanisms for detecting and responding to stimuli. These behaviors are not random but intentional, as organisms seek to maximize survival by optimizing their response to changing environmental conditions.

The evolution of toxins, venoms, and poisons is intimately connected to the selective pressures faced by organisms in their environments. From bacterial defense systems to venomous reptilian and insect strategies, these agents act as biochemical weapons against predators, competitors, and prey (Adler & Wüster, 2006). For instance, bacteria evolved toxins like diphtheria and tetanus to protect themselves against predation by other microorganisms and immune system responses. Similarly, many plants and animals have developed toxins as a means of defense against herbivores and other potential threats (Jiang et al., 2015).

Toxins evolve in response to environmental pressures and interactions with other species. Through processes of natural selection, organisms producing effective toxins that disrupt vital biological systems in their targets are more likely to survive. For example, the neurotoxic properties of snake venom evolved to rapidly incapacitate prey by targeting ion channels and disrupting neural transmission (Kini, 2008). Production of bacterial toxins can be regulated by quorum-sensing elements on the bacterial surface. Downregulation of pertussis toxin production and other virulence genes in *Bordetella pertussis* and other gram-negative bacteria was shown by sensing high concentrations of sulphate and other nutrients externally or, in turn, by upregulating cysteine desulfinase gene expression that increases internal concentrations of sulphate, which is also sensed by the bacterium as the nutrients in the growth environment change (Bogdan et. al., 2001).

While toxins have generally evolved to target specific molecules, they frequently cause unintended "off-target" effects. These off-target interactions can result in damage to a variety of organisms, including humans, and may lead to more serious health consequences than the intended effect. For example, botulinum neurotoxin, although primarily targeting acetylcholine release in the neuromuscular junction, can have systemic effects if improperly administered (Blasi et al., 1993).

Roles of Toxins and Venoms

Bacterial toxins, such as those produced by *Clostridium botulinum* (botulinum toxin) or *Corynebacterium diphtheriae* (diphtheria toxin), have played pivotal roles in affecting human mortality. These toxins are molecular products of bacterial metabolism that serve as weapons against hosts. The *botulinum* neurotoxin, for instance, targets the nervous system, blocking acetylcholine release, and leading to paralysis (Schantz & Johnson, 1999). The evolution of these toxins provided a significant survival advantage for the bacterium, allowing it to thrive in hostile environments like dry nutrient deplete soils. Other toxins like Staphylococcal enterotoxins are synthesized in the gut but have lesser effects on digestion than other enteric toxins, but more consequential proinflammatory effects mediated by the immune system and the overall immune response, leading to detrimental systemic effects that make them an attractive candidate for a bioweapon (Ahanotou et. al., 2006).

Marine organisms like cone snails (*Conus spp.*) and certain types of fish produce potent neurotoxins, such as conotoxins, which serve both predatory and defensive purposes. These toxins act on ion channels in the nervous system, causing paralysis in prey (Olivera et al., 1990). Similarly, plant toxins like ricin, found in the castor bean, represent biochemical agents that have evolved to deter ingestion of plants as food. Ricin inhibits protein synthesis by inactivating ribosomes, which can lead to organ failure if ingested or inhaled (Smith, 2010).

Venoms produced by reptiles and insects, such as the venom of the black widow spider (*Latrodectus mactans*) or rattlesnakes (*Crotalus spp.*), are sophisticated biochemical mixtures that act on the nervous and cardiovascular systems of their targets (Gordon & Mebs,

2000). These venoms have evolved to incapacitate prey or deter predators, providing the organism with both offensive and defensive tools for survival.

Therapeutic and Vaccines Applications

The development of vaccines marked a pivotal moment in humanity's fight against bacterial toxins. Vaccines against diphtheria, tetanus, and pertussis, for example, utilize detoxified versions of the toxins (toxoids) to stimulate the immune system and provide immunity without causing disease (Plotkin et al., 2013). The invention of these vaccines revolutionized public health and drastically reduced mortality from these bacterial infections.

In addition to vaccines, therapeutic antibodies, like those used to treat botulism or other toxin exposures, have become essential in the practice of modern medicine. Hyper-immune immunoglobulin, derived from the blood of individuals who have recovered from botulinum exposure, provides passive immunity and is used to treat poisoning (Shapiro et al., 2001). Development of a licensed heptavalent anti-toxin to treat seven serotypes most likely to cause human disease is produced by hyper-immunizing horses with toxins, purifying and de-speciating/removing the horse component of the purified immunoglobulin to produce Botulinum anti-toxin (BAT). BAT is available through the United States Strategic National Stockpile in emergencies to treat food-borne botulism and weaponized toxins (Gottron & Wyatt, 2023). To treat infantile botulism, a polyclonal antibody product derived from adults immunized with a serotype A and B vaccine is available. The United States Military is developing a highly potent heat-stable six-monoclonal-antibody formulation that can be administered by intramuscular injection to rapidly protect and/or treat warfighters before or after exposure to serotype A and B-derived neurotoxin in theatre (Tomic et. al., 2019).

Botox, derived from botulinum neurotoxin (BoNT), is an example of how a dangerous toxin has been harnessed for medical and cosmetic purposes. In low doses, BoNT can be used to reduce wrinkles, alleviate chronic migraines, and treat cervical dystonia. The therapeutic use of Botox highlights how understanding the molecular basis of toxins has led to beneficial medical applications (Carruthers & Carruthers, 2003). However, the same toxin, in larger doses, remains lethal, demonstrating the fine line between therapeutic benefit and harm. While BoNTs are already used to treat various medical conditions such as dystonia, spasticity, chronic migraines, and overactive bladder, research is continuously uncovering new potential applications. These include novel uses for pain management (e.g., in dentistry), gastrointestinal disorders, and even psychiatric conditions such as depression and anxiety, as BoNTs have been shown to impact neurotransmitter release and mood regulation (Demchenko et al. 2024; Dressler and Johnson 2022; Monesh et. al, 2025). Recent studies have shown its versatile effects, as Botox has been shown to reverse angiogenesis in the eyes of mice (Gregg et. al., 2025) and in humans to treat macular degeneration when administered in a manner to diminish "crow's feet" (Borodic, 2023).

Small molecule inhibitors of bacterial toxins have been difficult to identify, which has created a significant gap in the development of therapeutics to treat toxin exposure. Current treatments rely on immunoglobulins and monoclonal antibodies. A 60 kDa protein, bovine fetuin (BF), contains a glycosidic moiety that pertussis toxin binds. Concern about vaccine preparations that may possibly risk contamination by material associated with bovine spongiform encephalopathy led investigators to replace BF in the purification process for an acellular pertussis vaccine. They discovered a seven-amino-acid peptide that mimics the glycosidic moiety on BF that inhibits the biological activity of the toxin. The molecular structure of pertussis toxin bound to the peptide was determined. This peptide may represent a starting point for the design and synthesis of a bioactive small-molecule therapeutic to inhibit the toxin. Advances in AI/M-based algorithms, modeling programs, and computational analysis tools may further assist in its discovery (Bogdan et. al., 2003).

Development of individual toxoid vaccines for diphtheria (D), tetanus (T), and pertussis (P), and combined DTP vaccines was the first step in the 20th century to reduce significant morbidity and mortality in infants and adults from severe and frequently fatal disease. A safer acellular pertussis-containing version, DTaP, has evolved that contains select individual components rather than the whole inactivated *Bordetella pertussis* bacterium itself.

The development of protein-polysaccharide conjugation technology in the 1980s allowed the availability of novel vaccines against *Haemophilus influenzae* type b and serogroups of *Neisseria meningitidis* that have demonstrated very good safety and tolerability, together with the capability of eliciting a strong immunogenicity, combined with the demonstration of anamnestic antibody responses. The main advantages of the conjugation technology used in bacterial vaccines, due to the generation of a T cell-dependent immune response, are briefly outlined:

i. Improvement of the priming: immunogenicity in infants and young children (Ab-response, predominantly of the IgG1 isotype).
ii. Capability of eliciting an immunogenic memory response (production of long-lived memory B-cells) and a booster effect upon new contact with the specific antigen (revaccination).
iii. Capability of leading to affinity maturation of the Ab-response, with a consequent increased Ab-ag fit and improved opsonizing function.
iv. Generation of a mucosal immune response (secretory IgA and mucosally active IgG).
v. Reduction of the mucosal carriage (a prerequisite of herd protection).

Since the implementation of the Hib conjugate vaccines and their successful introduction into the pediatric immunization programs of some countries in the early 1990s has resulted in the near elimination of Hib meningitis, the accomplishment clearly led to the development of a series of equally successful experiences against other bacterial species relevant to global public health. The demonstration of the effectiveness of these immunization programs in children with a new generation of vaccines was the direct result of their exceptional immunological characteristics. The implementation of safe and effective meningococcal type C (Men C) vaccines to reduce endemic disease in children in the early 2000's occurred in the United Kingdom when three companies implemented their Men C vaccines. Next, a heptavalent pneumococcal conjugate vaccine (PCV-7) in the mid-2000s to reduce pneumococcal disease in children. Further advances in meningococcal and pneumococcal vaccines expanded the serotype coverage (i.e., Men AC, Men ACW135Y, PCV10, PCV13 globally (Borrow et. al., 2009 and Durando et. al., 2017) and more recently PCV 20 licensed for administration to children from 6 weeks to 17 years of age and recommended for adults over the age of 50 years of age (Shirley, 2023). Development of meningococcal type B (Men B vaccine) to prevent the most predominant strain causing meningitis in the US (Schillie et. al., 2024) and a pentavalent Men ABCW135Y to provide strain cross-protection will continue to increase coverage and reduce the incidence of meningococcal meningitis (Jenco, 2025).

Impact on Human Consciousness, Recognition, and Co-Existence

Before the 20th century, bacterial toxins such as those causing diphtheria, tetanus, and botulism were significant contributors to human morbidity and mortality (Graham et al., 2016).

Similarly, exposure to plant toxins, such as those found in poison ivy and ricin, led to widespread poisonings. The mechanisms of these toxins were often not understood, but their lethal effects on the human body were well recognized.

Certain toxins, such as those from hallucinogenic plants or neurotoxic venoms, can directly alter human consciousness. For example, the use of psychedelic substances like those derived from *LSD* or *psilocybin* has been shown to affect brain networks associated with sensory processing and perception (Carhart-Harris et al., 2016). Similarly, neurotoxins from insects and reptiles may cause profound changes in consciousness, from paralysis to altered mental states, as they affect the central nervous system (Gordon & Mebs, 2000). Synthetic Ziconotide (Prialt®), derived from the magical cone marine snail, is an approved analgesic to treat chronic pain in the conscious state, while others, like synthetic tirofiban derived from the saw scaled viper snake, are used to treat acute coronary syndrome, and synthetic lixisenatide derived from is used to treat Type II diabetes mellitus (Bordon et. al., 2020).

Humans have coexisted with toxic agents for millennia. Throughout history, humans have learned to both avoid and harness the power of these natural compounds. Whether through the development of antidotes or the use of toxins or toxoid-containing vaccines for medical purposes, humans' awareness of these agents has evolved into a sophisticated understanding that balances danger with utility (Harborne, 1993). The ability to coexist with these powerful biological agents, using them to protect life or to heal, speaks to humanity's capacity to adapt and thrive in a world full of potentially lethal substances.

Discussion

Ongoing research in the toxins field has revealed a wealth of potential therapeutic applications for toxins. For example, venom from the Brazilian pit viper (*Bothrops asper*) is being studied for its potential to treat heart disease, while other toxins are being explored as potential cancer therapies (Gutiérrez et al., 2016). Additionally, the use of venom-derived peptides to treat chronic pain is gaining attention in the pharmaceutical industry (Dussor & Xie, 2016). In addition, studies on inactivated forms of C3 cobra toxin protein, PVX and MMX, have revealed that they are powerful host-targeting therapies with the potential to treat viral infections and cancer in companion animals and humans (Yourist et. al., 1983; Bogdan et. al., 2025).

While the therapeutic use of toxins presents exciting opportunities, it is also recognized as a potential to foster ethical concerns and debate. The potential misuse of toxins in bioterrorism, for example, remains a significant concern (Feldmann & Geisbert, 2011). The safety of certain therapies, such as the use of botulinum toxin in aesthetic treatments, must be carefully regulated to prevent adverse outcomes (Benson et al., 2009). Patients receiving injections should be more aware of the potential for adverse events. Recently, an example of traditional food borne occurred when improperly prepared beans used in salads prepared and delivered widely in Russia led to a widespread outbreak of cases (Dorner, Martin B., 2025). The inappropriate administration of Botox for aesthetic enhancement has recently led to serious cases of overdose and hospitalizations when received at in-home "Botox parties" in 2024 (Dawson et. al., 2024). The United States Department of Defense and the Health and Human Services Office maintain and fund active programs to develop and/or supply vaccines and medical countermeasures kept in the United States Strategic National Stockpile to prevent and treat intentional exposure. This includes a plan that stockpiles ventilators for rapid distribution across the country in case of an intentional attack in populated areas (Gottron & Wyatt, 2023).

Conclusion

Cognitive biology is the discipline in which organisms recognize the processes that shape our world. Toxins, venoms, and poisons are natural agents that have evolved in response to selective pressures and serve as essential tools for survival. By recognizing environmental stimuli and reacting to them, they have evolved to adapt to and coexist with their nearest neighbors and harvest food sources for growth and survival. While dangerous, these agents offer tremendous insights into the mechanisms of life, death, and evolution. Through recognizing and understanding their roles and effects, humans have developed vaccines and therapeutic treatments that not only will save lives but also will trigger new possibilities for future medical use. As we continue to coexist with these powerful biological agents, our increasing understanding of their mechanisms of action will be crucial in shaping their medical use and protective measures across the world for generations to come. Combining licensed therapeutics and vaccines with over-the-counter drugs and nutraceuticals, and implementing ayurvedic Hindu medical approaches like diet, herbal treatment, and relaxation therapy through yoga may be of benefit against human exposure to toxins and venoms or other environmental poisons. This could positively impact health outcomes, shifting the balance towards life rather than death.

References

Adler, K., & Wüster, W. (2006). *The evolution of venomous organisms and the significance of toxinology.* BioScience, 56(5), 440-448.

Ahanotu, E., Alvelo-Ceron, D., Ravita, T., & Gaunt, E. (2006). *Staphylococcal Enterotoxin B as a Biological Weapon: Recognition, Management, and Surveillance of Staphylococcal Enterotoxin. Applied Biosafety*, 11(3) pp. 120-126.

Benson, P. M., Whitaker, J. S., & Bolding, P. (2009). *The safety profile of botulinum toxin for aesthetic use.* Dermatologic Surgery, 35(5), 804-809.

Blasi, J., Chapman, E. R., & Yamasaki, S. (1993). *Botulinum neurotoxins: Mechanism of action and structure-function relationships. Journal of Clinical Investigation*, 91(3), 1113-1120.

Bogdan, J. A., Alexander, P, Nazario-Larrieu, et al. (2001). *Bordetella Pertussis Auto-Regulates Pertussis Toxin Production through the Metabolism of Cysteine, Infection and Immunity*, 69:6823-6830.

Bogdan, J. A., Yuan, W., Long-Rowe, et al. (2003). *Identification of Peptides That Mimic the Pertussis Toxin Binding Site on Bovine Fetuin. Applied and Environmental Microbiology*, Vol. 69.

Bogdan, J. A., Yourist, J., Chaar, W., & Somers, D. (2025). *PVX, A Host-Targeting Broad-Spectrum Antiviral Therapeutic Derived from Cobra Venom Toxin. The 3rd International Conference on Toxins*. September 10-12, 2025.

Bordon, K., Takeno, C., Fornari-Baldo, E., et. al. (2020) *From Animal Poisons and Venoms to Medicines: Achievements, Challenges and Perspectives in Drug Discovery, Frontiers in Pharmacology*, Translational Pharmacology Section.

Borodic, G. (2023). *Botulinum toxin type A in multimodal management of age-related macular degeneration and related diseases, Toxicon,* Volume 236.

Borrow, R., & Findlow, J. (2009). *Prevention of meningococcal serogroup C disease by NeisVac-C, Expert Reviews Vaccines*, Mar;8(3):265-79.

Calow, P. (1998). *Handbook of functional morphology.* Wiley-Blackwell.

Carruthers, J. D., & Carruthers, A. (2003). *Botulinum toxin type A: Historical perspective and current indications. Plastic and Reconstructive Surgery*, 111(3), 818-823.

Carhart-Harris, R. L., et al. (2016). *The entropic brain: A theory of conscious states informed by neuroimaging research with psilocybin. Frontiers in Human Neuroscience*, 10, 293.

Dawson, E. F., Pateriya, S., Blashinsky, Z., & Landa-Galindez, E. (2024). *From Botox Party to Botulism: A Case Report and Public Health Warning on Adulterated Botulinum Toxin Injection Cureus, part of Springer Nature.* The Florida Medical Student Research Publications Open Access.

Dorner, M. (2025*). An outbreak of foodborne botulism caused by commercially canned mushrooms from Russia., Fachsymposium Lebensmittelmikrobiologie.*

Durando, P., Faust, S. & Torres, A. (2015). *Editorial: Immunological Features and Clinical Benefits of Conjugate Vaccines against Bacteria, Journal of Immunology Research.*

Dussor, G., & Xie, J. Y. (2016). *Venom peptides and pain research: Future directions. Trends in Pharmacological Sciences*, 37(5), 331-344.

Feldmann, H., & Geisbert, T. W. (2011). *Ebola and Marburg viruses: Pathogenesis and development of therapeutics. Current Opinion in Virology*, 1(3), 223-229.

Gordon, D., & Mebs, D. (2000). *The evolution and medical significance of venomous organisms. Nature Reviews Neuroscience*, 1(4), 254-260.

Gottron, F.; & Wyatt, T. (2023). *Congressional Research Service (CRS) Reports and Issue Briefs.*

Graham, D. Y., et al. (2016). *History of toxin-mediated human diseases. Clinical Microbiology Reviews*, 29(4), 757-785.

Harborne, J. B. (1993). *Introduction to Ecological Biochemistry.* Academic Press.

Jenco, M. (2025). *HHS approves pentavalent meningococcal vaccine for adolescents, young adults. American Academy of Pediatric News: Infectious Diseases, Vaccine/Immunization.* July 2.

Jiang, Y., et. al. (2015). *Evolution and molecular biology of plant toxins. Annual Review of Plant Biology*, 66, 419-442.

Kini, R. M. (2008). *Venom: The powerful biologically active molecules in snake venom and their therapeutic applications. Journal of the National Institute of Health*, 110(1), 1-12.

Monash, A., Tam, J., Rosen, O., et al. (2025). *Botulinum Neurotoxins: History, Mechanism, and Applications, A Review. Journal of Neurochemistry*, 169.

Olivera, B. M., et al. (1990). *Conotoxins: The toxins of marine snails. Trends in Pharmacological Sciences*, 11(4), 208-213.

Plotkin, S. A., et al. (2013). *Vaccines*. Elsevier Health Sciences.

Schantz, E. J., & Johnson, E. A. (1999). *Botulinum toxin: Clinical aspects. Clinical Microbiology Reviews*, 12(2), 43-56.

Schillie, S., Loehr, J., Chen, W. et. al. (2024) New *Dosing Interval and Schedule for the Bexsero MenB-4C Vaccine: Updated Recommendations of the Advisory Committee on Immunization Practices - United States, October, MMWR* Vol. 73 No. 49.

Shapiro, R. L., et al. (2001). *Botulism: Clinical aspects, epidemiology, and treatment. Clinical Infectious Diseases*, 33(7), 1343-1350.

Shirley, M. (2023). *20-Valent Pneumococcal Conjugate Vaccine: Pediatric First Approval. Paediatr Drugs*. Sep;25(5):613-619.

Smith, L. M. (2010). *The biological properties and medical uses of ricin and other plant toxins. Phytochemistry Reviews*, 9(3), 345-353.

Tomic, Milan T., Espinoza, Y. et. al. (2019) *Monoclonal Antibody Combinations Prevent Serotype A and Serotype B*

Inhalational Botulism in a Guinea Pig Model. Toxins (Basel), Apr 6;11(4):208.

Yourist, J., Haines, H.G., & Miller. K. D. (1983) *Inhibition of Herpes Simplex Virus Replication by Cobra Neurotoxoid, Journal of General Virology* 64. (Pt 7) (JUL):1475-81.

Molecular Target-Based Development of Drug-Like and Fluorescent Molecules for Alzheimer's Disease

Abha Sharma

Department of Medicinal Chemistry, National Institute of Pharmaceutical Education and Research
Raebareli, India

Abstract

Alzheimer's disease (AD) is a prevalent neurodegenerative disorder that poses multiple challenges for disease diagnosis and treatment. It is characterized by multiple pathological processes, including amyloid-beta (Aβ) accumulation, neurofibrillary tangles (NFTs), disruptions of neuronal and synaptic functions, and neuroinflammation, which serve as targets for the treatment of the disease as well as biomarkers. Biomarker-based assays are one of the important tools for the early diagnosis of AD. In recent years, researchers have pursued the development of theranostic probes for multi-target analysis of AD biomarkers. Herein, we work on various target based drug like molecules which includes natural biologically active scaffolds like flavone, coumarin, pyridoxal-based, and others, including benzothiazole, benzimidazole, benzothiadiazole, Imidazo pyridine etc. for the treatment of AD, like Cholinesterase inhibitors, A*β* aggregation inhibitors, GSK-3□ inhibitors, reactive oxygen species etc., fluorescent probes for detection of biomarkers, including analysis of A*β* (NFTs, A*β* aggregates, or A*β* oligomers), AChE, pH and viscosity, various proteins and multi-modal imaging and theranostic probes targeting A*β*, AChE. The diagnostic and treatment / theranostic ability of the designed molecules was assessed using in vitro assays like Ellman's assay and Thioflavin-T-based fluorescence assay, followed by in vivo evaluation of potent drug-like molecules.

Keywords: Alzheimer's disease, Fluorescent, Amyloid beta, Acetylcholinesterase, pH

Introduction

Alzheimer's disease (AD) is the most challenging neurodegenerative disease affecting mostly the older population above 65 years of age. More than 50 million are suffering globally, and the number is likely to get hiked up to 150 million by the year 2050 (Christina et al., 2018). Several hypotheses, for instance, the cholinergic hypothesis, the amyloid-β cascade hypothesis, metal dyshomeostasis, tau hypothesis, and oxidative stress, are some of the identified etiologies involved in pathogenesis. Damage to cholinergic neurons leads to cognitive decline, which has led to the development of acetylcholinesterase (AChE) inhibitors, providing symptomatic relief to AD patients. However, this age-related disease poses a problem of uncontrollable bioburden of metal ions in the brain. The metal dyshomeostasis, in turn, induces oxidative stress and the formation of amyloid-β (Aβ) plaques. Increase in bio-metals amplifies reactive oxygen species (ROS), which includes augmentation of lipid peroxidation, protein oxidation, nitration, and glycoloxidation. Oxidative stress aggravates the pathophysiological conditions of neuronal damage (Mecocci & Polidori, 2012). A molecule having additional antioxidant potency along with AChE inhibitory potential and metal chelating ability can be developed as a potential anti-AD agent. Studies have stated that peripheral anionic site (PAS) binding of AChE may promote Aβ aggregation. For PAS binding, aromatic rings with pi-pi stacking are recommended as derived from the docking studies with donepezil (Silva et al., 2020).

The multifactorial nature of AD with several interconnected etiologies led to the idea of developing multi-target directed ligands (MTDLs). The design of MTDL comprises two or more pharmacophore moieties, which are responsible for targeting multiple targets that are accountable for the pathogenesis of the disease. Therefore, MTDLs are single molecules with multiple functions that are strategically designed to serve the purpose of developing potential new chemical entities that can be used against AD (Pathak et al., 2024).

Compounds for Designing MTDL

Vitamin B6 (Pyridoxine) is known to reduce homocysteine levels in Alzheimer's patients, which leads to oxidative stress that is also involved in the formation of amyloid beta aggregates. The pyridoxine structure consists of hydroxyl groups, which may be helpful in metal chelation along with antioxidant properties (Zhuo et al., 2010 & Hashim et al., 2011). Pyridoxine-Resveratrol derivatives have been reported as pyridoxine-based compounds against AD (Pawar et al., 2023). Therefore, pyridoxine-triazole and pyridoxine carbamates were synthesized, characterized, and tested for acetylcholinesterase activity (Figure 1).

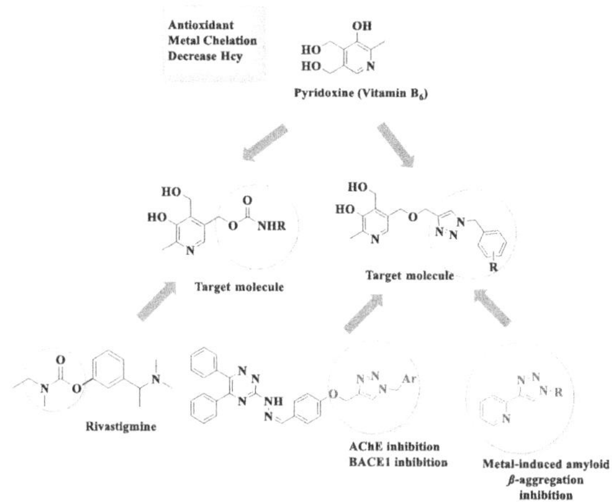

Figure 1: Design Strategy of MTDL based on pyridoxine

The Imidazo [1,2-a] pyridine moiety is the most important and exhibits anti-inflammatory, antiviral, antitumor, and antipyretic activities*. There are several drugs containing Imidazo [1,2-a] pyridine moiety on the market for the treatment of various diseases (Vanda et al., 2019).*

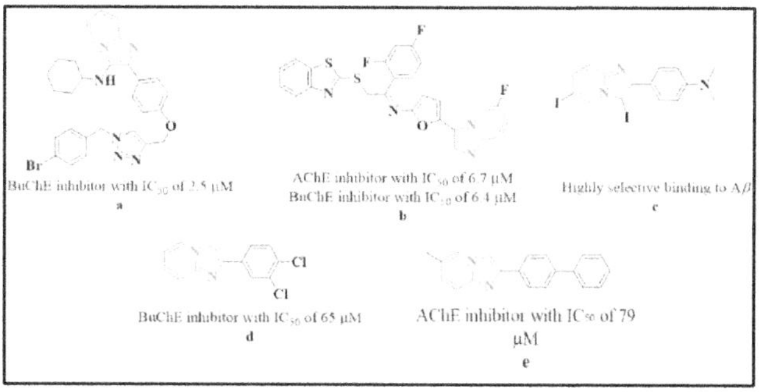

Figure 2. *Imidazo [1,2-a] pyridine containing fluorescent probes for AD detection and as anti-Alzheimer agents*

Rafaqat et al. have designed and synthesized imidazopyridine-based benzothiazole-oxadiazole hybrid derivatives (Hussain et al., 2024). The lead candidate was found to inhibit both AChE and BuChE with an IC_{50} of 6.7 □M and 6.4 □M. Huey et al. has designed and synthesized Imidazo [1,2-a] pyridine-based derivatives, which showed the strongest AChE inhibition with IC_{50} of 79 □M and

BuChE inhibition with IC_{50} of 65 □M (Kwong et al., 2019). Therefore, pyrone-fused imidazo[1,2-a]pyridine derivatives were synthesized, characterized, and tested for cholinesterase inhibitory activity.

Chromones are heterocyclic molecule that contains oxygen and are widely present in fungi and plants. Chromones and their derivatives have a variety of pharmacological properties, including the capacity to inhibit several enzymes and to exhibit antioxidant, antifungal, antibacterial, anticancer, antiviral, antihypertensive, and anti-inflammatory effects (Maicheen et al., 2023; Gaspar et al., 2024). Maicheen *et al.* synthesized novel chromone derivatives using the modified Baker Venkataraman reaction, and their antioxidant activity was compared to Quercetin. Compound 2e and 2j (figure-3) showed potent antioxidant activity in *in vitro* antioxidant assay (Maicheen et al., 2023). Similarly, S. Abdpour *et al.* developed a series of molecules containing chromone and pyridinium. When compared to donepezil (IC_{50} = 0.023 and 3.4 µM), the standard drug, the majority of the molecules had excellent BuChE inhibition action (IC_{50} = 1.9–0.006 µM) and good AChE inhibition (IC_{50} = 9.8–0.71 µM) (Abdpour et al., 2021). Therefore, chromone derivatives were synthesized, characterized, and tested for cholinesterase inhibition activity and antioxidant potential.

Fluorescent probes are emerging tools in multiple fields such as biomedical research, clinical diagnostics, analytical chemistry (Yang et al., 2020, Kong et al., 2020, Mojban et al., 2020) etc. Fluorescent probes designed to selectively bind biomarker amyloid beta (Aβ) aggregates developing for the diagnosis of AD. Reported dyes for Aβ aggregate detection, including Congo Red, Chrysamine G, and Thioflavin (Figure 3). Among these, Th-T is one of the most widely used fluorescent stains for labelling Aβ fibrils and neurofibrillary tangles in postmortem brain tissues obtained from AD patients (Bao et al., 2021; Lv et al., 2020). Among the various radio iodinated probes reported to date, an imidazo [1,2-a] pyridine-based iodinated molecule has demonstrated particularly promising in vitro and in vivo characteristics (Newberg 2006).

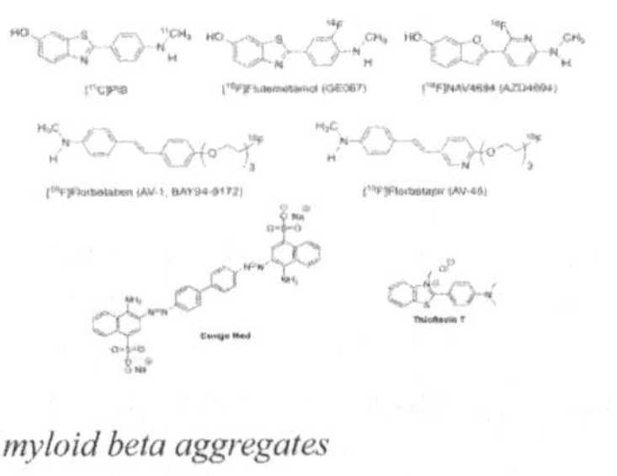

Figure 3. *Fluorescent Probes for sensing Amyloid beta aggregates*

pH is usually diverse in the biological system, ranging from neutral, slightly basic, to strongly acidic. Lysosomes (pH 4.0–5.0) are acidic, the cytoplasm (pH 7.2) is almost neutral, and mitochondria (pH ~8) are slightly alkaline.2,3 Any abnormality in these pH ranges can agitate the cellular functions that can result in the diseased condition. Some examples of abnormal pH leading to serious cellular dysfunction and protein degradation are tumorous cells having lower intracellular pH values than the normal extracellular pH (Parolini et al., 2009). Also, the alteration in lysosomal pH values has been linked with neurodegenerative disorders (Holopainen et al., 2001). The fluorescent probes containing various scaffolds such as BODIPY, cyanine, fluorescein, rhodamine, etc., are reported as sensors for pH. These probes possess good photophysical properties but lack in disadvantages include, such as the phenolic proton in fluorescein severely affecting the fluorescent properties, though it is tuned in the later analogs with halogen atoms (Lin et al., 1999). Similarly, carbocyanine dyes were linked with aniline and amino substituents to introduce pH-dependent properties (Myochin et al., 2011). However, these fluorophores have an inherent property to photo-bleach that could interfere with bioimaging applications. We have synthesized a library of pH-sensitive photostable fluorescent probes and studied the effect of substituents on them to develop compounds to overcome some limitations (Chaudharan et al., 2025).

Conclusion

The small-molecule-based drug-like and fluorescent compounds were designed for targeting multiple targets of complex AD. The selection of molecules was inspired by natural products such as pyridoxine, chromone, and other scaffolds that have druglike properties. The longer wavelength emissive fluorescent molecules were designed, synthesized, and tested for sensing biomarkers of AD.

Acknowledgement

Thankful to the Department of Pharmaceutical, Ministry of Chemicals and Fertilizers, and the Director of NIPER-R for providing facilities.

References

Abdpour, S.; Jalili-Baleh, L.; Nadri, H.; et al. (2021). Chromone Derivatives Bearing Pyridinium Moiety as Multi-Target-Directed Ligands against Alzheimer's Disease. Bioorg Chem, 110.

An, Y., Li, S., Huang, X., Chen, X., Shan, H., & Zhang, M. (2022). The role of copper homeostasis in brain disease. International journal of molecular sciences, 23(22), 13850.

Bao, W., Xie, F., Zuo, C., Guan, Y., & Huang, Y. H. (2021). PET neuroimaging of Alzheimer's disease: radiotracers and their utility in clinical research. Front Aging Neurosci, 13: 624330.

Cao, Y.; Liu, X.; Zhang, J.; Liu, Z.; et al. (2023). Design of a Coumarin-Based Fluorescent Probe for Efficient In Vivo Imaging of Amyloid-beta Plaques. ACS Chem. Neurosci, 14, 829–838.

Chaudhran, P. A., Mastoli, S. Y., Dey, A., et al. (2025). Dual-State Emissive and Substituent-Tunable pH-Sensitive Bis-Heterocyclic Fluorescent Probes. The Journal of Organic Chemistry, 90(16), 5359-5371.

Chen, X., Li, Y., Kang, J., et al. (2023). Application of a novel coumarin-derivative near-infrared fluorescence probe to amyloid-β imaging and inhibition in Alzheimer's disease. Journal of Luminescence, 256, 119661.

Christina, P. (2018). Alzheimer's Dis. Int. World Alzheimer Rep., 1–48.

Gaspar, A.; Garrido, E. M. P. J.; et al. (2024). Biological and Medicinal Properties of Natural Chromones and Chromanones. ACS Omega. American Chemical Society, 21706–21726.

Gunnlaugsson, T., Leonard, J. P., & Murray, N. S. (2004). Highly selective colorimetric naked-eye Cu (II) detection using an azobenzene chemosensor. Organic letters, 6(10), 1557-1560.

Hashim, A.; Wanga, L.; Junej, K.; et al. (2011). Vitamin B6s Inhibit Oxidative Stress Caused by Alzheimer's Disease-Related CuII-B-Amyloid Complexes- Cooperative Action of Phospho-Moiety. Bioorganic Med. Chem. Lett, 21 (21), 6430–6432.

Hien, N. K., Bay, M. V., et al. (2020). Coumarin-based dual chemosensor for colorimetric and fluorescent detection of Cu2+ in water media. ACS Omega, 5(33), 21241-21249.

Holopainen, J. M.; Saarikoski, J.; et al. (2001). Elevated lysosomal pH in neuronal ceroid lipofuscinoses (NCLs). Eur. J. Biochem, 268, 5851.

Hu, H., Yao, Y., Deng, L., Wang, H., & Xue, L. (2024). Synthesis and Properties of a Triphenylamine-Based Fluorescent Probe for the Continuous Detection of Cu2+ and EDTA. Chemistry Select, 9(12), e202304968.

Huang, X., Li, J., Guo, Y., et al. (2025). Ultrafast detection of bisulfite by a unique quinolinium-based fluorescent probe and its applications in smartphone-assisted food detection and bioimaging. Talanta, 282, 126977.

Kong, J.; Zhang, J.; Wang, Y.; et al. (2020). Bioinspired pH-Sensitive Fluorescent Peptidyl Nanoparticles for Cell Imaging. ACS Appl. Mater. Interfaces, 12, 4212.

Lee, M. H.; Han, J. H.; et al. (2013). Two-Color Probe to Monitor a Wide Range of pH Values in Cells. Angew. Chem., Int. Ed., 52, 6206.

Liao, Y. C., Venkatesan, P., et al. (2016). A coumarin-based fluorescent probe for thiols and its application in cell imaging. Sensors and Actuators B: Chemical, 232, 732-737.

Lin, H. J.; Szmacinski, H.; Lakowicz, J. R. (1999). Lifetime-based pH sensors: indicators for acidic environments. Anal. Biochem., 269, 162.

Lin, T. W., & Huang, S. D. (2021). Direct and simultaneous determination of copper, chromium, aluminium, and manganese in urine with a multielement graphite furnace atomic absorption spectrometer. Analytical Chemistry, 73(17), 4319-4325.

Lv, Z., Li, L., Man, Z., et al. (2020). Polarity-activated super-resolution imaging probe for the formation and morphology of amyloid fibrils. Nano Research, 13 (9), 2556-2563.

Maicheen, C., Churnthammakarn, C., et al. (2023). One-Pot Synthesis and Evaluation of Antioxidative Stress and Anticancer Properties of an Active Chromone Derivative. Molecules, 28 (7).

Mecocci, P., Polidori, M. C. (2012). Antioxidant Clinical Trials in Mild Cognitive Impairment and Alzheimer's Disease. Biochim. Biophys. Acta - Mol. Basis Dis., 1822 (5), 631–638.

Mojbafan, M., Zanjanchi, F., Taherkhani, M. (2020). The influence of physical factors on the halochromic behavior of the pH-sensitive sulfonphthaleine dyes: a DFT study. Chem. Pap., 74, 4355.

Myochin, T., Kiyose, K., Hanaoka, K., et al. (2011). Rational Design of Ratiometric Near-Infrared Fluorescent pH Probes with Various pKa Values, Based on Aminocyanine. J. Am. Chem. Soc., 133, 3401.

Newberg, A. B., Wintering, N. A., et al. (2006). Safety, biodistribution, and dosimetry of 123I-IMPY: a novel amyloid plaque-imaging agent for the diagnosis of Alzheimer's disease. J Nucl Med, 47 (5), 748-54.

Orioli, R., Belluti, F., Gobbi, S., Rampa, A., & Bisi, A. (2024). Naturally Inspired Coumarin Derivatives in Alzheimer's Disease Drug Discovery: Latest Advances and Current Challenges. Molecules, 29(15).

Parolini, I.; Federici, C.; Raggi, C.; et al. (2009). Microenvironmental pH is a Key Factor for Exosome Traffic in Tumor Cells. J. Biol. Chem., 284, 34211.

Pathak, C., & Kabra, U. D. (2024). A comprehensive review of multi-target directed ligands in the treatment of Alzheimer's disease. Bioorganic Chemistry, 144, 107152.

Pawar, R., Chaudhran, P., et al. (2023). Chemical modifications of pyridoxine for biological applications: an overview. Current topics in medicinal chemistry, 23(2), 98-113.

Shang, J. Y., Li, Y., Chen, K., & Li, H. (2021). Synthesis and properties of an AIE fluorescent probe for Cu2+ detection based on the ESIPT system. Chemical Papers, 75(5), 1851-1859.

Sheng, R., Wang, P., Gao, Y., et al. (2008). Colorimetric test kit for Cu2+ detection. Organic letters, 10(21), 5015-5018.

Silva, M. A.; Kiametis, A. S.; Treptow, W. (2020). Donepezil Inhibits Acetylcholinesterase via Multiple Binding Modes at Room Temperature. J. Chem. Inf. Model., 60(7), 3463-3471.

Strianese, M., Milione, S., Bertolasi, V., & Pellecchia, C. (2013). Iron and manganese pyridoxal-based complexes as fluorescent probes for nitrite and nitrate anions in aqueous solution. Inorganic Chemistry, 52(20), 11778-11786.

Tang, F., Wang, K., Liu, X., et al. (2023). Small Molecular Fluorescent Probes for Alzheimer's Disease-Associated Active Species. Chemistry – A European Journal, 29 (35), e202300592.

Trinh, N., Bhuskute, K. R., Varghese, N. R., et al. (2024). A Coumarin-Based Array for the Discrimination of Amyloids. ACS sensors, 9(2), 615-621.

Webb, B. A., Chimenti, M., Jacobson, M. P., Barber, D. L. (2011). Dysregulated pH: a perfect storm for cancer progression. Nat. Rev. Cancer, 11, 671.

Yang, X.-L., Wu, W.-X., et al. (2020). A facile strategy to construct a fluorescent pH-sensitive drug delivery vehicle. Polymer, 197, No. 122496.

Zhuo, J. M., Praticò, D. (2010). Acceleration of Brain Amyloidosis in an Alzheimer's Disease Mouse Model by a Folate, Vitamin B6 and B12-Deficient Diet. Exp. Gerontol., 45 (3), 195–201.

Cognition Beyond Brains: Redefining Intelligence in Nature

Dr. Shiv Satchit

Journalist and Media Technology Consultant, UK

Abstract

This article proposes a paradigm shift in understanding intelligence, moving away from brain-centric models towards a framework of scale-free cognition. Intelligence is positioned not as an exclusive product of complex nervous systems, but as a fundamental, decentralised property of biological life. By synthesising research from bioelectric, botany, microbiology, and physics, this paper explores how non-neural systems, including plants, microbes, and individual cells, process information, retain memories, and engage in goal-directed decision-making.

Key themes include decentralised biological intelligence, cellular cognition through bioelectric signalling, reconceptualising oncology as cognitive failure, the heart-brain connection, and universal informational patterns.

The paper concludes that recognising nature as a cognitive ecosystem necessitates a transition in medicine and science from mechanical interventions towards therapies that restore relational coherence within living systems.

Keywords: scale-free cognition, basal cognition, bioelectric signalling, distributed intelligence, plant intelligence, heart-brain connection

Introduction

Beyond the Neural Frontier

For over a century, Western medicine and biological science have operated under what can be termed the "boss myth" - the assumption that the brain serves as the sole command centre whilst the body's cells function merely as mindless labourers (Levin, 2021).

This brain-centric hierarchy has profoundly shaped our understanding of intelligence, cognition, and consciousness. However, emerging research in scale-free cognition and basal cognition is dismantling this paradigm, revealing that intelligence is not a privilege reserved for organisms with neurons but rather a fundamental property of biological systems themselves (Lyon et al., 2021).

This paradigm shift carries profound implications for clinical practice, environmental ethics, and our philosophical understanding of nature. If intelligence exists as a spectrum of information processing from subcellular biochemistry to cosmic-scale structures, then our approach to health, disease, and ecological stewardship requires fundamental reconsideration. This paper explores the evidence for cognition beyond brains and examines its implications for medicine, ethics, and human wellbeing.

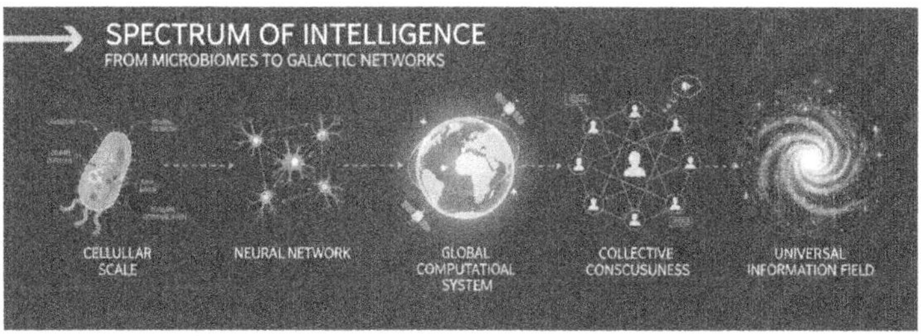

The Architecture of Intelligence

Deconstructing the Brain-Centric Bias

Traditional models of intelligence have equated cognitive capacity with neural complexity, centralised processing, and human-like reasoning. This anthropocentric view assumes a hierarchical 'ladder of intelligence,' with humans positioned at the apex (Baluška & Mancuso, 2009).

However, mounting evidence suggests that intelligence manifests through diverse strategies across biological systems, none inherently superior to others - merely adapted to different ecological niches.

Intelligence, in its most fundamental form, can be defined as the capacity to sense and respond to environmental conditions, solve problems adaptively, learn from experience, and make decisions that enhance survival (Levin, 2019). By this definition, intelligence pervades nature far beyond the boundaries of the nervous system.

The Spectrum of Biological Intelligence

NON-COGNITIVE INTELLIGENCE SPECTRUM		
Types of Intelligence	Core Description and Function	Examples of Manifestation
Emotional (EQ)	Perceiving, understanding, managing, and utilising one's own emotions and those of others.	Self-awareness, empathy, managing conflict, inspirational leadership.
Intuitive Intelligence	Non-conscious, rapid understanding based on pattern recognition, implicit learning, and somatic cues.	A 'gut feeling,' sudden insight, trusting instinct in complex decisions.
Bodily-Kinesthetic	Using one's body skillfully to solve problems, express ideas, or create.	Athletics, dance, surgery, craftsmanship, acting.
Enhanced access to intuition Intelligence.	A calm, coherent system is better at receiving and interpreting the subtle signals from your implicit knowledge and emotional memory.	A seasoned doctor, after quietly reviewing a complex case with puzzling symptoms, suddenly has a specific 'hunch' about a rare disease not yet tested for, leading to a crucial and correct diagnosis.
Coherent Intelligence	A harmonised state where heart rhythms, brain activity, and the nervous system are in sync, optimising perception and access to inner wisdom.	The 'flow state,' deep meditative clarity, heart-centered decision-making, peak performance. This coherent state is key to enhancing access to intuition. A calm, coherent system is

		better at receiving and interpreting the subtle signals from your implicit knowledge and emotional memory.
Distributed Intelligence	Intelligence shared by trees and spread across a forest environment. **Mycorrhizal Networks:** This 'wood wide web' consists of intricate underground fungal networks connecting the roots of multiple trees and plants	**Bird Flocks & Fish Schools** bird/fish stay close, avoid collision, match speed for safety and navigation without a leader. Trees share water, nitrogen, phosphorus, and carbon with saplings or injured trees, provide defence signals (warning chemicals for pests or disease), and even recognise their own kin to direct more resources toward them [1, 2].
Collective Intelligence (CI)	Shared or group intelligence that emerges from the collaboration, competition, or consensus of many individuals.	**CI** used for social insects like ants or bees, or human groups, implies a unified decision-making process or problem-solving that emerges from the interaction of many agents working toward a conscious goal.
Hereditary/Genetic Intelligence	Intelligence isn't written purely in your genes, nor is it only about your environment. It is deeply intertwined. Intelligence is genetically influenced, but not in a simplistic manner.	No intelligence genes have been found makes you smart or not. Instead, it looks like thousands of genes each contribute tiny amounts to your cognitive abilities; if you inherited genes from bright parents, you are born with potential.
Consciousness (as Substrate)	The fundamental capacity for awareness and experience; the 'ground' that allows information to be integrated into a unified reality.	Subjective experience, self-awareness, the binding of senses into a single perception.
Cosmic / Universal Intelligence	The observable principle of self-organisation, order, and emergent complexity inherent in the universe itself. The 'Highest Order of Systemic Integration.'	If a cell's health depends on its connection to the organ and the organ's health depends on its link to the body, then its health must depend on its connection to the wider world (meaning, purpose, and the infinite).

Cellular Cognition and the TAME Framework

Dr Michael Levin's "Technological Approach to Mind Everywhere" (TAME) framework provides a systematic method for identifying and measuring cognition at multiple scales (Levin, 2019). TAME defines cognition as the ability of a system to pursue goals through diverse means, emphasising teleology - goal-directedness as a measurable biological reality rather than anthropomorphic projection.

During salamander limb regeneration, individual cells work collectively towards a specific anatomical goal state, ceasing proliferation only when the correct morphology is achieved (Mondia et al., 2011). This demonstrates that cells possess inherent problem-solving capabilities and can navigate 'morphospace', the landscape of possible body forms to reach predetermined endpoints.

Levin's concept of "cognitive light cones" maps the scale of an organism's intelligence based on the spatial and temporal complexity of goals it can pursue. A single cell operates within a limited light cone, whilst multicellular organisms coordinate across larger scales, and ecosystems demonstrate goal-directed behaviour across even greater dimensions (Levin, 2019).

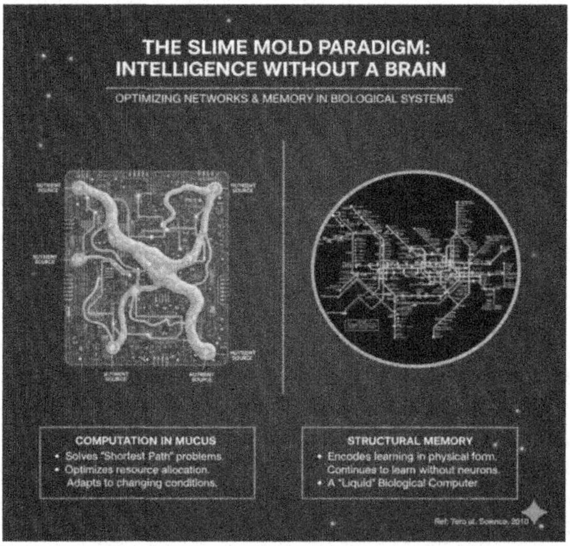

1. Unicellular Cognition

Perhaps most remarkably, intelligence manifests in organisms entirely lacking nervous systems. The slime mould *Physarum polycephalum* navigates mazes, optimises network designs comparable to human-engineered transport systems, anticipates periodic events, and demonstrates spatial memory - all without a brain or neurons (Nakagaki et al., 2000).

When researchers recreated the geography around Tokyo metro using food sources, *Physarum* designed a network nearly identical to the city's rail system, suggesting inherent computational sophistication at the cellular level (Tero et al., 2010).

2. Plant Cognition

Plants demonstrate sophisticated cognitive abilities despite lacking centralised nervous systems. Research has documented that mimosa plants exhibit habituation, learning to ignore repeated harmless stimuli and retaining this memory for weeks (Gagliano et al., 2016). The Venus flytrap (*Dionaea muscipula*) employs a counting mechanism, requiring two trigger hair stimulations within twenty seconds before expending energy to close its trap -an elegant solution to the problem of false alarms (Böhm et al., 2016).

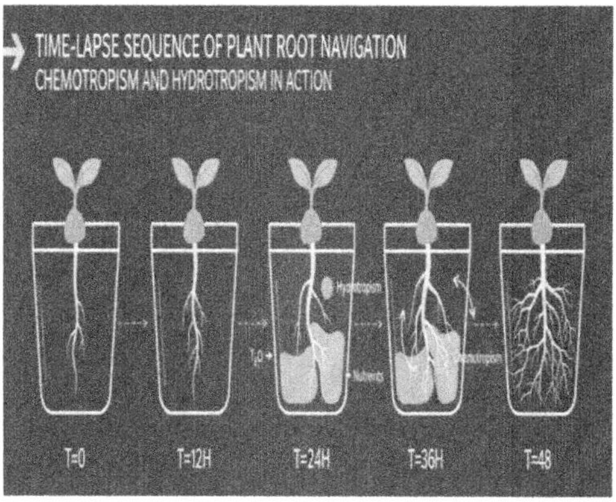

Beyond individual plant intelligence, forest ecosystems function as distributed cognitive networks. Simard's (2021) research on mycorrhizal networks, popularly termed the "Wood Wide Web," reveals that trees actively share resources through underground fungal connections.

Mature "mother trees" transfer carbon, water, and nutrients to struggling seedlings, whilst chemical signalling coordinates forest-wide responses to pest attacks. This resource sharing occurs preferentially amongst kin, suggesting recognition and strategic decision-making capabilities (Simard et al., 2015).

The Cosmic Web and Fractal Architecture

Vazza and Feletti (2020) conducted a quantitative comparison between the neuronal network of the human brain and the cosmic web of the universe, revealing structural similarities approaching 99 per cent. Both systems organise matter into filamentous networks with comparable spectral densities and information content.

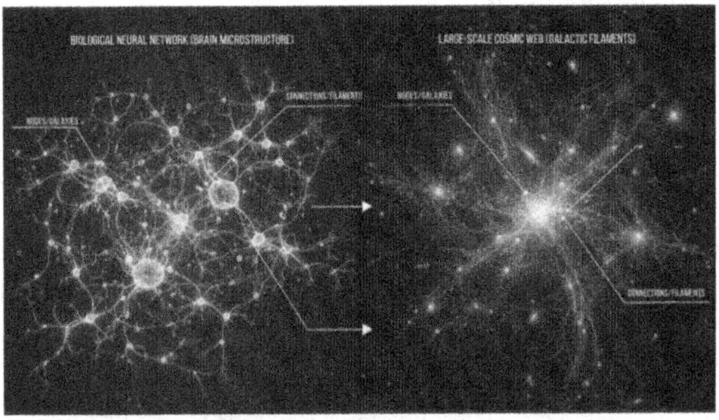

This fractal architecture—the repetition of patterns across scales—suggests that nature employs similar informational strategies from the subcellular to the cosmic level. The mycorrhizal networks of forests, the neural networks of brains, and the filamentary structure of galaxies all reflect this fundamental organisational principle.

3. Immune Intelligence

Your immune system thinks without your brain:

- **Pattern recognition:** Distinguishes "self" from "non-self"
- **Learning & memory:** Remembers pathogens for decades (vaccination)
- **Adaptation:** Generates novel antibodies for new threats
- **Decision-making:** Determines appropriate responses (inflammation, tolerance)

This is **embodied cognition:** Intelligence distributed throughout biological systems.
Here's something even more personal: your own body contains multiple intelligences that operate independently of your brain. Your immune system is perhaps the most sophisticated example. It faces an incredible challenge; it must distinguish 'self' from 'non-self' among trillions of cells and countless foreign molecules. And it must do this without any input from your conscious mind.

The immune system learns and remembers. When you get vaccinated, immune cells encounter a pathogen, develop a response, and create memory cells that can persist for decades. This is why you don't get chickenpox twice, because your immune system remembers and responds faster the second time. The system adapts continuously, generating millions of different antibody variations through genetic recombination, essentially running evolutionary experiments inside your body to find the best match for new threats.

And it makes nuanced decisions: attack this bacterium, tolerate this food protein, destroy this cancer cell, don't attack this fetus during pregnancy. These decisions happen through complex biochemical networks - it's distributed cognition without a central processor. This is embodied intelligence, and it reminds us that thinking doesn't only happen in our heads.

4. Collective Intelligence

Social insects provide compelling examples of distributed cognition, where complex problem-solving emerges from simple individual behaviours without centralised control. Honeybee colonies solve optimisation problems comparable to the travelling salesman problem when determining efficient foraging routes (Lihoreau et al., 2012).

Ant colonies cultivate fungus gardens, maintain temperature regulation, and coordinate defence strategies - all without any individual ant comprehending the colony's overall strategy (Hölldobler & Wilson, 2009).

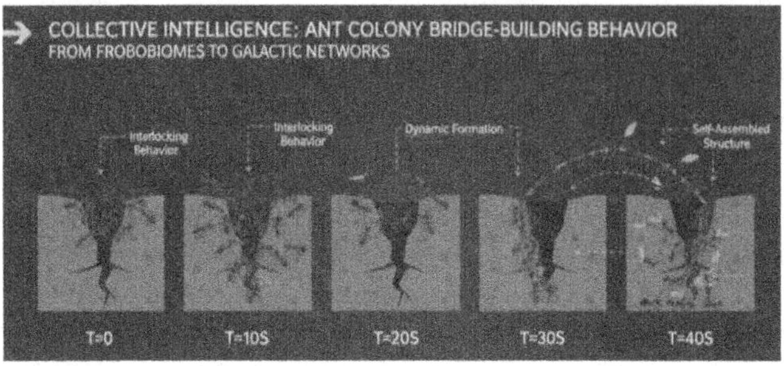

Bird flocks demonstrate emergent intelligence through murmurations, where thousands of individuals create coordinated patterns following simple local rules with no central coordinator (Reynolds, 1987). These collective behaviours reveal that intelligence can transcend individual minds, arising from the interactions between components of a system.

Bacteria engage in quorum sensing, a form of collective decision-making where population density influences gene expression and behaviour (Waters & Bassler, 2005). This demonstrates that even unicellular organisms process information about their social environment and modify behaviour accordingly.

5. Distributive Intelligence

Distributed Intelligence	Intelligence shared by trees and spread across a forest environment. **Mycorrhizal Networks:** This 'wood wide web' consists of intricate underground fungal networks connecting the roots of multiple trees and plants.	**Bird Flocks & Fish Schools** bird/fish stay close, avoid collision, match speed for safety and navigation without a leader. Trees share water, nitrogen, phosphorus, and carbon with saplings or injured trees, provide defence signals (warning chemicals for pests or disease), and even recognise their own kin to direct more resources toward them [1, 2].

6. Universal Patterns: Intelligence in "Inanimate" Systems

Structural Intelligence in Non-Living Systems

The definition of intelligence need not be confined to biological systems. Consider the piezoelectric properties of bone: mechanical stress generates electrical signals that direct bone growth and remodelling (Bassett & Becker, 1962). This represents a form of distributed computation where the material itself processes mechanical information and responds adaptively.

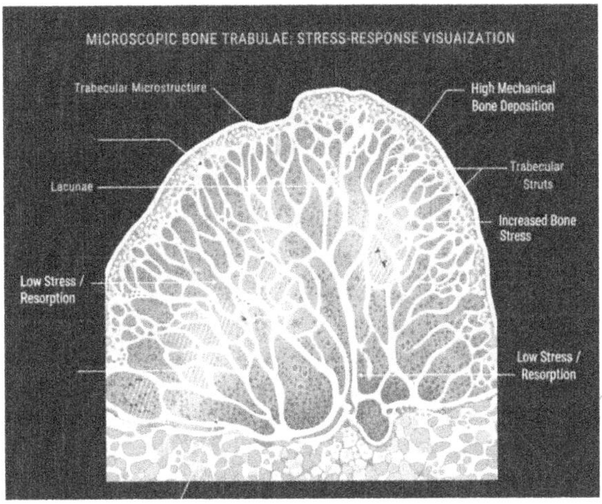

Rivers demonstrate "fluid logic" in solving energy dissipation problems through meandering patterns that minimise energy expenditure (Leopold & Langbein, 1962). Whilst not cognitive in the biological sense, such systems exhibit optimisation principles analogous to problem-solving.

The Sentience of the Natural World

Plant Communication and Stress Responses

Recent research from Tel Aviv University documented that plants emit ultrasonic sounds when under stress—frequencies of 20-100 kilohertz produced when cut, dehydrated, or otherwise damaged (Khait et al., 2023). Whilst inaudible to humans without equipment, these acoustic emissions could potentially be detected by nearby organisms, suggesting a previously unrecognised communication channel.

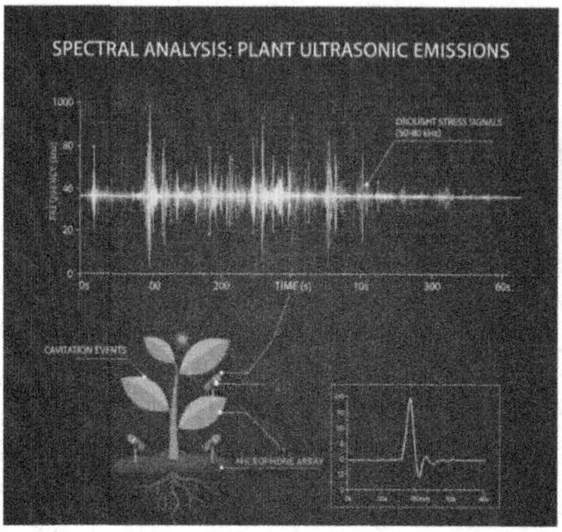

Plants respond to vibrational stimuli through phototropism—directional growth influenced by sound. Gagliano et al. (2012) demonstrated that plant roots grow preferentially towards sounds associated with water, even when moisture is not present, indicating anticipatory behaviour based on acoustic cues.

Interspecies Coherence

The concept of interspecies coherence suggests that the electromagnetic fields generated by different organisms can interact and potentially synchronise (Hagens, 2010). Given that the heart's electromagnetic field extends beyond the body, and plants possess their own bioelectric fields, the possibility emerges of resonance between human emotional states and plant physiology.

Whilst rigorous research in this area remains limited, preliminary studies suggest that human intent and emotional state may influence plant growth and health through mechanisms not yet fully understood (Tompkins & Bird, 1973; Radin et al., 2008). This phenomenon, colloquially termed the "King Charles Effect" in reference to the then-Prince Charles's practice of speaking to plants, warrants further scientific investigation through controlled, replicated studies.

Bioelectric Signalling and Cellular Agency

The Bioelectric Software of Life

Every cell maintains a membrane potential - a voltage gradient across its membrane. Far from being metabolic noise, these bioelectric signals constitute a high-speed information network (Levin, 2021).

When cells communicate via gap junctions, they establish what Levin terms a "social contract," coordinating to build specific organs and maintain tissue identity. This bioelectric signalling system operates faster than biochemical processes, providing rapid, spatially organised information crucial for development and regeneration (Levin, 2014).

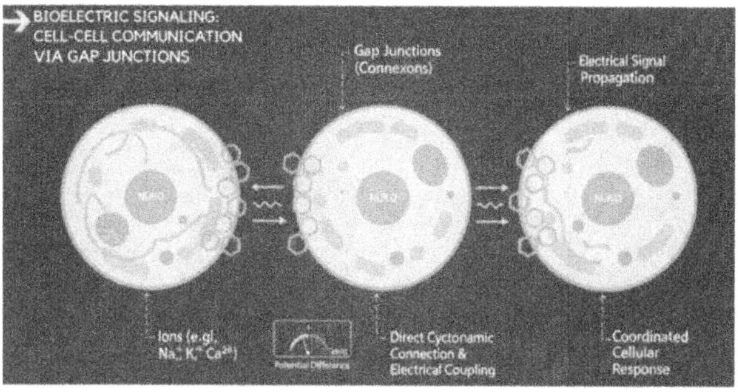

Research on planarian flatworms demonstrates the profound role of bioelectric patterns. When decapitated, planarians regenerate complete heads with memories intact, suggesting that memory storage is distributed throughout the body rather than localised exclusively in neural tissue (Shomrat & Levin, 2013). This challenges the conventional understanding of memory as purely neurological.

Implications for Oncology

Viewing cancer through the lens of cellular cognition reframes oncology fundamentally. In this framework, tumours represent not merely genetic mutations but cognitive breakdowns—cells that have "forgotten" their role within the multicellular collective and reverted to unicellular, resource-hoarding behaviour (Levin, 2021). The cancer cell becomes a "lonely cell," severed from the bioelectric communication network that maintains tissue identity.

This perspective suggests therapeutic approaches beyond genetic targeting. Bioelectric modulation could potentially "re-educate" cancer cells, convincing them to rejoin the collective rather than simply destroying them (Chernet & Levin, 2013). Such approaches would treat cancer as a failure of communication rather than purely as a mutation problem.

The Heart as a Cognitive Centre

The Heart-Brain Connection

Conventional cardiology views the heart primarily as a mechanical pump. However, neurocardiology research reveals the heart as a sophisticated information-processing centre with its own intrinsic nervous system - the "heart brain", containing approximately 40,000 to 50,000 neurons (Armour, 2004). This cardiac nervous system can sense, learn, remember, and make decisions independently of the cranial brain.

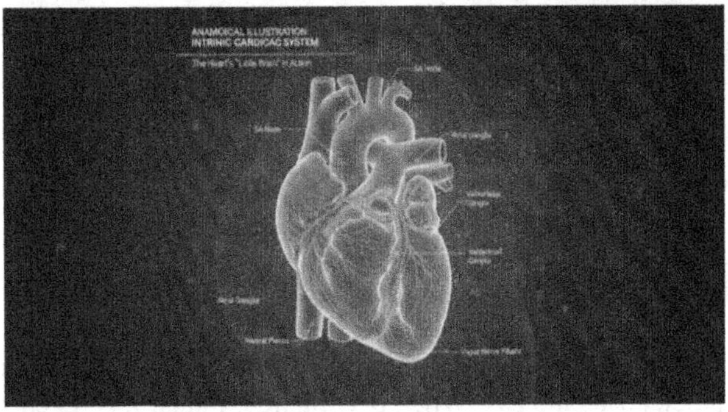

The heart-brain dialogue operates through four primary pathways (McCraty, 2015):

1) **Neurological Communication**: The vagus nerve contains approximately 80% afferent fibres, meaning the majority of neural traffic flows from heart to brain rather than the reverse. These ascending signals significantly influence higher cognitive functions, emotional processing, and decision-making.
2) **Biochemical Communication**: The heart functions as an endocrine gland, producing hormones including atrial natriuretic peptide (ANP), which regulates blood pressure, and oxytocin, the "bonding hormone," in concentrations comparable to brain production (Gutkowska & Jankowski, 2012).
3) **Biophysical Communication**: The heart's pressure wave, travelling faster than blood flow itself, provides rhythmic synchronisation signals to every cell. Baroreceptor activity influences electrical activity in the thalamus and cortex (Lacey & Lacey, 1978).

4) **Electromagnetic Communication**: The heart generates the strongest rhythmic electromagnetic field in the body—approximately 60 times greater in amplitude than brain electrical activity and over 100 times stronger magnetically. This field extends several feet from the body and carries emotional information (McCraty et al., 2009).

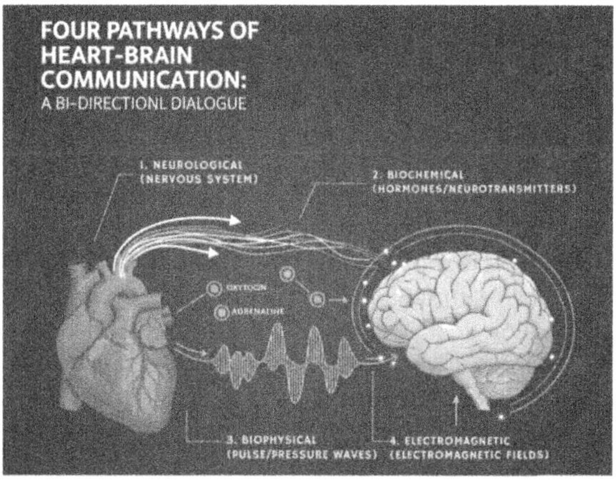

Heart Rate Variability and Coherence

Heart rate variability (HRV)—the variation in time intervals between heartbeats—reflects the dynamic interplay between sympathetic and parasympathetic nervous system activity. Coherent HRV patterns, characterised by smooth, sine-wave-like rhythms, correlate with improved cognitive performance, emotional stability, and physiological efficiency (McCraty & Zayas, 2014).

When the heart operates in a coherent state, it facilitates cortical function; when chaotic, it inhibits cognitive processing. This suggests the heart serves not merely as a responsive organ but as a primary influence on brain function and whole-body physiology (McCraty, 2015).

Implications for Medicine and Health

From Mechanical to Ecological Medicine

Recognizing the body as a cognitive ecosystem, not a machine, necessitates therapeutic approaches that restore communication and coherence, not simply repair or replace parts. The "Wood Wide Web" model, where resource sharing ensures collective

survival, provides a framework for understanding health as fundamentally relational.
Clinical protocols should prioritise:

1) **Heart-brain coherence training**: Given that vagal afferents dominate heart-brain communication, techniques that establish cardiac coherence—such as HRV biofeedback and coherent breathing—should become standard interventions for supporting healing states (Lehrer & Gevirtz, 2014).
2) **Bioelectric diagnostics**: Measuring cellular voltage gradients and bioelectric patterns could provide early detection of pathological states before structural changes become evident on conventional imaging (Levin, 2021).
3) **Practitioner coherence**: If electromagnetic fields carry emotional information and influence nearby biological systems, clinician emotional state may affect patient physiology. Healthcare providers should cultivate coherent internal states to optimise therapeutic environments (McCraty et al., 2009).

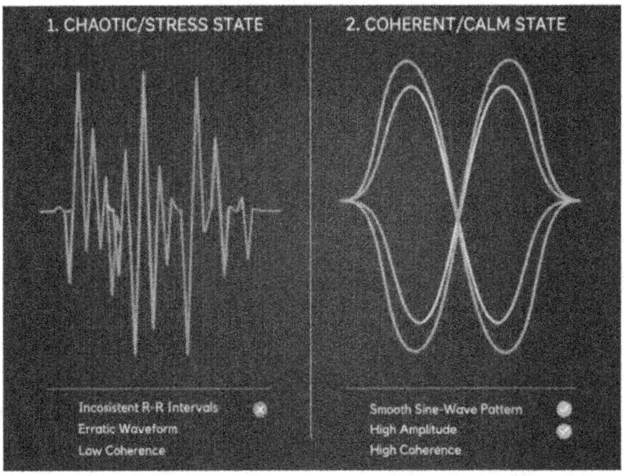

Regenerative Medicine and Cellular Reprogramming

Understanding basal cognition opens possibilities for regenerative medicine beyond conventional approaches. Rather than micromanaging molecular pathways, therapies could engage the inherent decision-making capacities of cells (Levin, 2019). Research on xenobots (living robots constructed from frog cells that self-organise and navigate environments without genetic programming) demonstrates that cells possess remarkable autonomous problem-solving capabilities (Kriegman et al., 2020).

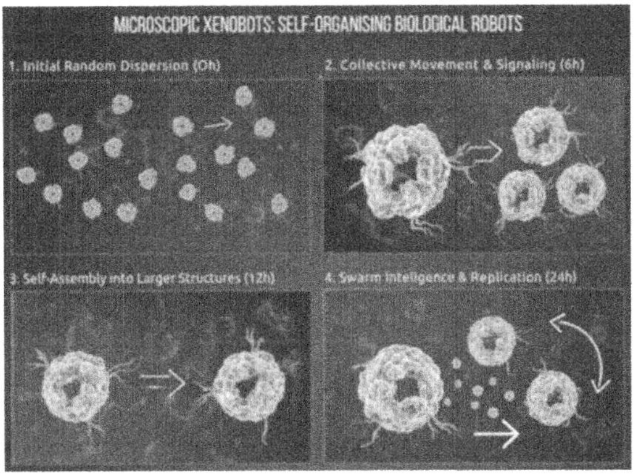

If salamander cells can collectively navigate morphospace to regenerate perfect limbs, the question becomes not whether human cells possess similar capabilities, but how to unlock them. Bioelectric manipulation may provide the key to triggering regenerative programmes dormant in human tissues.

Environmental and Ethical Implications

Expanding the Moral Circle

If forests think and communicate, if plants make decisions and remember experiences, and if individual cells demonstrate agency, then our ethical frameworks require expansion. Environmental destruction becomes not merely resource depletion but the silencing of cognitive entities and the disruption of intelligent ecosystems (Gagliano, 2017).

Indigenous wisdom traditions have long recognised the intelligence and sentience of non-human nature. Scientific validation of these perspectives suggests that traditional ecological knowledge offers crucial insights for sustainable human-nature relationships (Kimmerer, 2013).

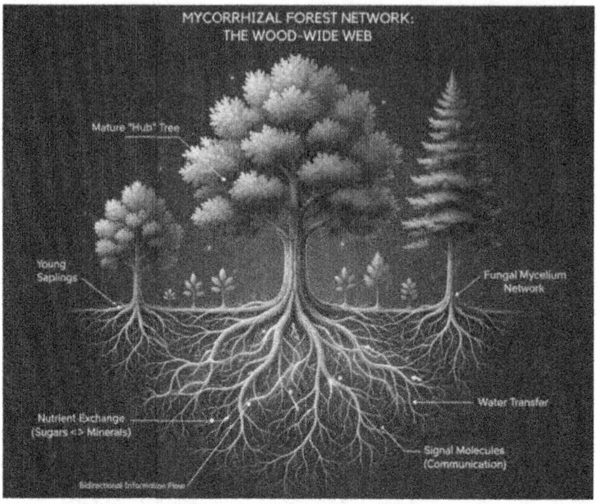

From Hierarchy to Network

The shift from viewing nature as a hierarchical "ladder of intelligence" to recognising a "web of intelligences" has profound implications. Rather than asking "How smart is it?" In comparison to humans, we must ask, "How does it think?" Each form of cognition is exquisitely adapted to its ecological context. An octopus's distributed intelligence, thinking through its arms, is not inferior to human cognition it is differently intelligent (Godfrey-Smith, 2016).

This perspective dissolves the boundary between self and environment. Humans are not separate from or superior to nature but rather participants in vast cognitive networks extending from cellular to ecological to potentially cosmic scales.

Practical Applications

Agriculture and Food Production

Understanding plant communication enables agricultural practices that work with natural intelligence rather than against it. Companion planting leverages plant-plant communication for pest management and improved growth (Barrios, 2007). Recognising soil as a cognitive ecosystem teeming with microbial intelligence and fungal networks suggests regenerative practices that support rather than deplete this biological computation.

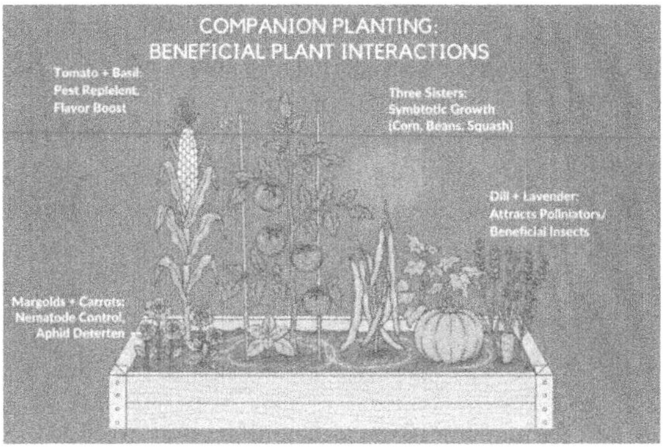

Biomimetic Technology

Swarm intelligence algorithms, inspired by collective behaviours of bees and ants, already optimise routing, scheduling, and resource allocation in technological systems (Bonabeau et al., 1999). Deeper understanding of cellular computation and bioelectric signalling could inspire novel computing architectures and artificial intelligence designs fundamentally different from current neural network models.

Conservation and Ecosystem Management

Preserving ecosystems means protecting not merely species diversity but cognitive networks. When a forest is clear-cut, we destroy not only trees but an intelligent communication system that has developed over centuries. Conservation strategies must account for the cognitive interdependence of ecosystem components (Simard, 2021).

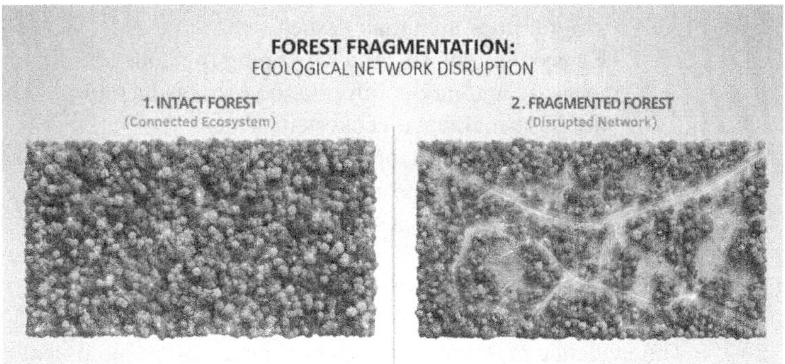

Critical Evaluation and Future Directions

Methodological Considerations

Whilst evidence for cognition beyond brains accumulates, methodological challenges remain. Anthropomorphism must be carefully distinguished from legitimate attribution of cognitive capacities. Rigorous operational definitions of terms like 'intelligence,' 'decision-making,' and "memory" across diverse biological systems are essential (Sims, 2020).

Experimental designs must account for alternative explanations of observed behaviours. For instance, plant responses to environmental stimuli might reflect evolved mechanisms rather than active cognition. However, as Trewavas (2017) argues, the distinction between hardwired responses and cognitive processing becomes increasingly difficult to maintain as we understand the sophistication of plant signalling.

Future Research Priorities

Several areas warrant intensive investigation:

1) **Bioelectric cartography**: Mapping bioelectric patterns across development, disease, and healing to establish diagnostic and therapeutic applications.
2) **Interspecies field interactions**: Rigorous, well-controlled studies examining whether human electromagnetic fields influence plant or animal physiology under experimental conditions.
3) **Cognitive regeneration**: Systematic exploration of how bioelectric manipulation can trigger regenerative programmes in human tissues.
4) **Ecosystem cognition**: Developing methods to measure and model information processing and decision-making at ecosystem scales.
5) **Philosophical frameworks**: Refined conceptual tools for discussing consciousness, sentience, and moral status across the spectrum of biological systems.

Conclusion: Towards a Sentient Symphony

The evidence presented marks a profound shift in biological understanding, from viewing the brain as the exclusive seat of intelligence to recognising cognition as a fundamental property woven throughout biological systems. Intelligence manifests across scales: from slime mould memory and mycorrhizal resource sharing to cardiac programming of brain function and the piezoelectric computation within bone.

This paradigm reveals the universe not as a collection of inert matter but as a symphony of sentient systems operating through universal informational architectures. When plants emit ultrasonic distress signals and human hearts broadcast emotional data through electromagnetic fields 5,000 times stronger than brain waves, the boundary between self and environment dissolves. We become participants in continuous bioelectric conversations with everything around us.

For medicine, this realisation is transformative. If cancer represents cognitive failure of cellular communication, healing becomes not mechanical repair but restoration of coherence. By shifting focus from genetic hardware to bioelectric software and recognising the role of emotional intent, we transition from mechanics to gardeners of living ecosystems.

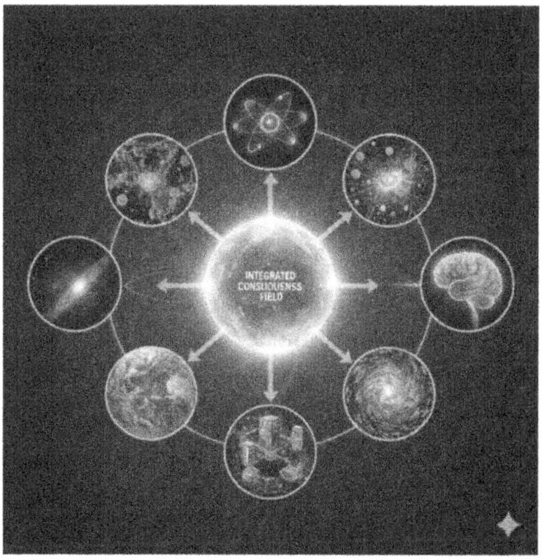

Intelligence emerges as the fundamental drive of the universe to move from chaos to order—whether through tree roots, neurons, or mycorrhizal networks, life constantly seeks connection, cooperation, and meaning. To heal the body, we must honour the intelligence residing in every cell and acknowledge our profound entanglement with the sentient world.

A Path Forward

The transition from viewing nature as an inanimate resource to recognising it as a sentient partner represents perhaps the most important evolution in human consciousness since the Scientific Revolution. This shift moves us from being mechanics of dead machines to participants in living minds.

For healthcare professionals, this means:

- Acknowledging every cell as an active cognitive agent
- Treating the heart-centred emotional state as the primary gateway to physiological recovery
- Validating the interconnection between human bioelectric fields and environmental influences
- Adopting therapeutic approaches that restore relational coherence

The "universal mind" is not a mystical abstraction but the observable, self-organising principle of nature. As we enter clinics and communities with this understanding, we do so with the humbling knowledge that we are not masters of nature but conscious expressions of its infinite intelligence. Recognition of distributed intelligence is no longer optional; it is a biological and survival imperative.

When humans view themselves as separate from nature, we mirror the behaviour of cancer cells, severing communication with the larger collective to focus on unchecked individual growth. Long-term human sustainability depends on re-establishing the "social contract" with nature, recognising that our health is fundamentally entangled with planetary bioelectric systems.

The future lies not in dominating nature but in learning from 3.8 billion years of evolutionary problem-solving. By acknowledging that non-cognitive does not mean non-intelligent, we access nature's vast research and development. We stop attempting to hack nature and begin apprenticing under it.

References

Armour, J. A. (2004). Cardiac neuronal hierarchy in health and disease. *American Journal of Physiology-Regulatory, Integrative and Comparative Physiology*, *287*(2), R262-R271.

Baluška, F., & Mancuso, S. (2009). Deep evolutionary origins of neurobiology: Turning the essence of 'neural' upside-down. *Communicative & Integrative Biology*, *2*(1), 60-65.

Barrios, E. (2007). Soil biota, ecosystem services and land productivity. *Ecological Economics*, *64*(2), 269-285.

Bassett, C. A., & Becker, R. O. (1962). Generation of electric potentials by bone in response to mechanical stress. *Science*, *137*(3535), 1063-1064.

Böhm, J., Scherzer, S., Krol, E., Kreuzer, I., et al. (2016). The Venus flytrap *Dionaea muscipula* counts prey-induced action potentials to induce sodium uptake. *Current Biology*, *26*(3), 286-295.

Bonabeau, E., Dorigo, M., & Theraulaz, G. (1999). *Swarm intelligence: From natural to artificial systems*. Oxford University Press.

Chernet, B. T., & Levin, M. (2013). Transmembrane voltage potential is an essential cellular parameter for the detection and control of tumor development in a *Xenopus* model. *Disease Models & Mechanisms*, *6*(3), 595-607.

Gagliano, M. (2017). The mind of plants: Thinking the unthinkable. *Communicative & Integrative Biology*, *10*(2), e1288333.

Gagliano, M., Renton, M., Depczynski, M., & Mancuso, S. (2012). Experience teaches plants to learn faster and forget slower in environments where it matters. *Oecologia*, *175*(1), 63-72.

Gagliano, M., Vyazovskiy, V. V., Borbély, A., et al. (2016). Learning by association in plants. *Scientific Reports*, *6*(1), 38427.

Godfrey-Smith, P. (2016). *Other minds: The octopus, the sea, and the deep origins of consciousness*. Farrar, Straus and Giroux.

Gutkowska, J., & Jankowski, M. (2012). Oxytocin revisited: Its role in cardiovascular regulation. *Journal of Neuroendocrinology*, *24*(4), 599-608.

Hagens, S. (2010). *The geometry of meaning*. Gainesville, FL: Brown Walker Press.

Hölldobler, B., & Wilson, E. O. (2009). *The superorganism: The beauty, elegance, and strangeness of insect societies*. W. W. Norton & Company.

Khait, I., Lewin-Epstein, O., Sharon, R., et al. (2023). Sounds emitted by plants under stress are airborne and informative. *Cell, 186*(7), 1328-1336.

Kimmerer, R. W. (2013). *Braiding sweetgrass: Indigenous wisdom, scientific knowledge and the teachings of plants*. Milkweed Editions.

Kriegman, S., Blackiston, D., Levin, M., & Bongard, J. (2020). A scalable pipeline for designing reconfigurable organisms. *Proceedings of the National Academy of Sciences*, *117*(4), 1853-1859.

Lacey, B. C., & Lacey, J. I. (1978). Two-way communication between the heart and the brain: Significance of time within the cardiac cycle. *American Psychologist*, *33*(2), 99-113.

Lehrer, P. M., & Gevirtz, R. (2014). Heart rate variability biofeedback: How and why does it work? *Frontiers in Psychology*, *5*, 756.

Leopold, L. B., & Langbein, W. B. (1962). The concept of entropy in landscape evolution. *US Geological Survey Professional Paper*, *500*(A), A1-A20.

Levin, M. (2014). Molecular bioelectricity: How endogenous voltage potentials control cell behavior and instruct pattern regulation in vivo. *Molecular Biology of the Cell*, *25*(24), 3835-3850.

Levin, M. (2019). The computational boundary of a "self": Developmental bioelectricity drives multicellularity and scale-free cognition. *Frontiers in Psychology*, *10*, 2688.

Levin, M. (2021). Bioelectric signaling: Reprogrammable circuits underlying embryogenesis, regeneration, and cancer. *Cell*, *184*(8), 1971-1989.

Lihoreau, M., Chittka, L., & Raine, N. E. (2012). Trade-off between travel distance and prioritization of high-reward sites in traplining bumblebees. *Functional Ecology*, *26*(5), 1194-1204.

Lyon, P., Keijzer, F., Arendt, D., & Levin, M. (2021). Reframing cognition: Getting down to biological basics. *Philosophical Transactions of the Royal Society B*, *376*(1820), 20190750.

McCraty, R. (2015). *Science of the heart, Volume 2: Exploring the role of the heart in human performance*. Heart Math Institute.

McCraty, R., Atkinson, M., & Bradley, R. T. (2004). Electrophysiological evidence of intuition: Part 1. The surprising role of the heart. *Journal of Alternative and Complementary Medicine*, *10*(1), 133-143.

McCraty, R., Atkinson, M., Tomasino, D., & Bradley, R. T. (2009). The coherent heart: Heart-brain interactions, psychophysiological coherence, and the emergence of system-wide order. *Integral Review*, *5*(2), 10-115.

McCraty, R., & Zayas, M. A. (2014). Cardiac coherence, self-regulation, autonomic stability, and psychosocial well-being. *Frontiers in Psychology*, *5*, 1090.

Mondia, J. P., Levin, M., Omenetto, F. G., & Oné, D. L. (2011). Long-distance signals are required for morphogenesis of the regenerating *Xenopus* tadpole tail, as shown by femtosecond-laser ablation. *PLoS ONE*, *6*(10), e24953.

Nakagaki, T., Yamada, H., & Tóth, Á. (2000). Maze-solving by an amoeboid organism. *Nature*, *407*(6803), 470.

Radin, D., Hayssen, G., Emoto, M., & Kizu, T. (2008). Double-blind test of the effects of distant intention on water crystal formation. *Explore*, *2*(5), 408-411.

Reynolds, C. W. (1987). Flocks, herds and schools: A distributed behavioral model. *Computer Graphics*, *21*(4), 25-34.

Shomrat, T., & Levin, M. (2013). An automated training paradigm reveals long-term memory in planarians and its persistence through head regeneration. *Journal of Experimental Biology*, *216*(20), 3799-3810.

Simard, S. W. (2021). *Finding the mother tree: Discovering the wisdom of the forest.* Alfred A. Knopf.

Simard, S. W., Beiler, K. J., Bingham, M. A., et al. (2012). Mycorrhizal networks: Mechanisms, ecology and modelling. *Fungal Biology Reviews*, *26*(1), 39-60.

Sims, M. (2020). How to count biological minds: Symbiosis, the free energy principle, and reciprocal multiscale integration. *Synthese*, *199* (Suppl 3), 2157-2179.

Tero, A., Takagi, S., Saigusa, T., et al. (2010). Rules for biologically inspired adaptive network design. *Science*, *327*(5964), 439-442.

Tompkins, P., & Bird, C. (1973). *The secret life of plants.* Harper & Row.

Trewavas, A. (2017). The foundations of plant intelligence. *Interface Focus*, *7*(3), 20160098.

Vazza, F., & Feletti, A. (2020). The quantitative comparison between the neuronal network and the cosmic.

Ārṣa Theories of Cognition: Scriptural Foundations and Developmental Perspectives

Dr. Ashutosh Pareek

Professor, Samrat Prithviraj Chauhan Government College, Ajmer, Rajasthan

Abstract

The Ārṣa vision of holistic development provides a complete path for spiritual as well as cognitive awakening. Psychological well-being—with a blissful state of mind, soul, and sense organs—is an essential component of health (Sushruta *Sutra Sthana* 15/41). The Ārṣa scriptures establish cognitive development holistically, integrating intellect, morality, and spirituality—unlike Western models that focus purely on mechanics. They employ practices like yoga, meditation, *Anviksiki* (critical reasoning), and principles from SamkhyaYoga, emphasizing the inner self (*Atman*). These inner self experiences knowledge through stages like sense-mind, ego-mind, and intellect (*buddhi*), leading to self-realization. This contrasts with purely external, skill-based learning. The Panchakosha system outlines the internal and external dimensions of physical, spiritual, and emotional cognitive development. The Panchakosha theory from ancient Indian wisdom views well-being as five interconnected layers of existence: *Annamaya* (food/physical), *Pranamaya* (energy), *Manomaya* (mind/emotion), *Vijnanamaya* (wisdom/intellect), and *Anandamaya* (bliss). True health arises from balance and harmony across all layers, progressing from gross to subtle and guiding a holistic path to inner peace and fulfilment beyond mere physical health. The Panchakosha model provides a map for self-discovery, showing that genuine well-being comes from consciously balancing and aligning these five dimensions, ultimately revealing the innate joy within. By attending to the physical, vital, mental, intellectual, and spiritual dimensions of being, individuals can cultivate a balanced and integrated life characterized by deep fulfilment and joy. Yoga, meditation, the seven chakras, sixteen samskaras, four ashramas, five maha-yajnas, reincarnation, the karma system, and moksha are key features that fostered a cognitive approach in ancient times through the broader teachings of Ārṣa scriptures. This study explores the interconnections between motivational aspects of personality and psychological well-being.

Keywords: Ārṣa, Cognition, Dharma, Anvikshiki, Panch Koshas, Karma, Meditation.

Introduction

"Om. That is Whole; this is Whole.
From the Whole, the Whole arises.
When the Whole is taken from the
Whole, the Whole alone remains.
Om. Peace, Peace, Peace."

Interpretations of the famous Vedas or reactions to these ancient Indian scriptures that encompass psychology, religion, and philosophy are the key to cognitive development. Some applied parts of the respective writings are sometimes referred to in Western therapeutic approaches that deal with meditation (e.g., Kabat-Zinn, 1996; Wallace and Shapiro, 2006; Walsh and Shapiro, 2006). They also contain elements of brain development, spiritual wellbeing, and societal connectivity, so it is a fact that intelligence is not an exclusive byproduct of complex nervous systems but a fundamental decentralized property of all biological life. This paper seeks to explore the origins, core principles, and holistic approach to cognitive development through Ārṣa theories.

Ārṣa Foundation: Cognition Development

The "Ārṣa" system, also known as the "Rishi-Kula", is an ancient Indian social and cultural system based on the teachings and practices of the rishis (sages) and their descendants. It is deeply rooted in the Vedic tradition and has influenced various aspects of Human development.

"May your intentions be united; may your hearts be alike.
May your minds be in harmony, so that you may live together in
peace and mutual understanding.'

As per the hymns from the Rigveda, Vedic thoughts have inspired humans to move forward with co-existence, co-operation, and equality, avoiding conflicts. Ārṣa cognition refers to the infallible, doubt-free understanding attributed to ancient Rishis or divine sages in Indian philosophy. It transcends ordinary sensory perception, enabling direct insight into both material and super-sensuous realms. As per Ārṣa theory, cognition is not just brain activity; It is a whole-organism, environment-embedded phenomenon.

Nobel aspirant Prof. Amartya Sen said, "We live in the present, but that is a tiny bit of time; it passes as we talk. The current moment, as vivid as it is, does not tell us much about who we are, how we can reasonably see ourselves, and where we would place our loyalties if and when we face divisions. Our identities are strongly influenced by the past." [iii] The lack of cognition is described in the Mahabharata. As Duryodhana said to Krishna, who is not the god of the gods, do as you do for me? [iv]

O Krishna! I know what Dharma is, but I am unable to bring myself to follow it. I know what Adharma is, but I am unable to refrain from it. Vedas always seek intelligence. As per Athrvaveda 'Dhee', 'Medha' 'Sumati' are the different stages for cognition. Dhee (Wed) means power of learning, Medha (Perception) means power of retention, and Sumati (Wavec) means power of discrimination.[v] Vedic Medha reflects intellect, wisdom, or mental capacity, enabling a person to acquire knowledge and understand it accurately. In Vedic literature, Medha is given special importance as it reveals a person's mental and spiritual strength. It is considered a source of empowerment, helping individuals seek answers to life's questions and move toward the truth. Yajurveda says- "That wisdom (intelligence) which the hosts of gods and the ancestors revere and worship—by that wisdom, O Agni (Divine Fire), make me wise and endowed with understanding today. Svaha." [vi]

Oh, intellectual God, the type of supreme intelligence which was always worshipped and gained by the saints and our ancestors, provide us with the same intellectual intelligence. The Ācārya was expected to possess profound knowledge. So Maharshi Yaska said: "Who is an Ācārya (teacher)? An Ācārya teaches proper conduct, imparts knowledge and meaning, and cultivates understanding and wisdom." [vii]

The path of fearlessness is inherently aligned with the well-being of both the self and the world. This is why the path of fearlessness becomes a declaration of universal welfare alongside personal well-being. The Atharvaveda highlights this aspect of friendship-based fearlessness, stating:

"May I be free from fear from friends; may I be free from fear from foes.
May I be free from fear from those known to me and from those unknown.
May I be free from fear by night and free from fear by day.
May all directions and all hopes be friendly and auspicious to me." [viii]

Cognitive Development: Modern Concepts

As per modern psychologists, children progress through four universal stages, each with distinct cognitive abilities. Piaget's four stages are the most acceptable stages by the modern society.

1. **Sensorimotor Stage** (Birth to 2 years): At this stage, babies experience the world through senses such as sight, touch, sound, etc., and physical action (grabbing, crawling, mouthing, etc.).
2. **Preoperational Stage** (2 to 7 years): At this stage, a rapid development of language, imagination, and symbolic thinking is specifically observed. A child can use words and pictures but can't think logically yet.
3. **Concrete Operational Stage** (7 to 11 years): At this stage, a child can perform mental operations, like adding, subtracting and understanding cause and effect – but only with concrete objects or events they can see or touch. They understand concepts of conservation, reversibility, classification, and seriation. At this age, a child can think logically about real things, but not yet about abstract ideas.
4. **Formal Operational Stage** (12 years and up): At this stage, a child develops the ability to think abstractly and logically. A child can think about the ideas, possibilities, and hypothetical situations. A teenager can debate the pros and cons of the situation.

Jean Piaget[ix] showed the world that children are not miniature adults. They have their own ways of seeing and understanding. The stage-based theory helped us appreciate that learning is a gradual, active process, not just passive absorption.

Minimal Cognition is studied as sense-respond-adapt behavior. Even bacteria show chemotaxis, quorum sensing, and adaptive learning. In modern aspect Cognition was reframed as biological information processing, not consciousness. The cognitive cell as per bacterial behavior reconsidered by P Lyon.[x]

Scriptural Foundation of the Knowledge Cognition

The Rishis say that the Vedas are the Word of God. At the very beginning of the Creation, man received knowledge of the Vedas. The Rigveda is the most ancient of all the Holy Scriptures in the world. Knowledge of the Vedas is not intended for a particular race or country. God revealed this knowledge for the greatest good of mankind.

According to Vedanta, there are two categories of things that the mind is drawn to: Shreyas, or that which is beneficial and capable of bringing about enduring bliss, and Preyas, or that which is pleasant in the short term. A wise man weighs and makes a distinction between the excellent and the pleasurable when they come to a man.

"Both the Good (Śreyas) and the Pleasant (Preyas) approach a person.
The wise, after examining them, distinguish between the two.
The wise choose the Good over the Pleasant,
while the unwise choose the Pleasant for the sake of comfort, gain, and security."

Here is a beautiful stanza in the second chapter of the first cycle of Kathopanishad, in which it is said that the good and the pleasant come to a man, and the thoughtful mind turns around them and distinguishes. The wise choose out the good from the pleasant, but the dull soul chooses the pleasant rather than the getting of his good and its having.

Cognition emerges from elemental balance, not just neural firing. The cognitive study in Ārṣa scriptures is established through disciplined observation of mind and body. Yoga is a methodology for cognitive development, not just philosophy. Practices like Dhyana, Pratyahara, and Samadhi are systematically applied after cognition.

Cognition as a fundamental property of life

Indian thought does not see cognition as a late evolutionary product but as co-emergent with life itself. This resonates with the Indian idea that Chetana (Sentience) pervades all living beings (Sarva Bhuta). Upanishad said, "Consciousness is present in all beings" and "Indeed, everything in this universe is Brahman." [xiii] So as per Ārṣa scripture, minimal cognition can be seen as the earliest expression of Chetana manifesting through matter.

Unicellular Life: Anu Chetana

Ārṣa philosophy recognizes anu (the minute) as dynamically intelligent, not inert. Single-celled organisms sense, decide, avoid, harm, and seek nourishment. This clearly reflects Buddhi in its most primordial form, not conscious thought but directional intelligence. Katha Upanishad says, “The Self (Ātman) is subtler than the subtlest and greater than the greatest.” [xiv] Minimal cognition is buddhi at the scale of anu, intelligence without self-reflection.

Plants as living, responsive beings (Vanaspati Chetana)

Plants are not passive matter but living, responsive organisms with awareness. Ayurveda and ancient Indian culture treated plants as sentient participants in life, not resources. Modern plant neurobiology echoes this view through signaling, memory, and learning. Ayurveda said, “Plants are endowed with consciousness” and “Herbs grow upon the Earth.” [xv] Plant cognition reflects non-neural intelligence, expanding our definition of cognition itself.

Minimal Cognition and the Panch Mahabhutas

Cognition arises from elemental organization, not only from nervous systems. In unicellular life and plants, cognition is expressed through Ap (water), Agni (metabolism), and Vayu (movement). So, there is no nervous system required; only organized elemental interaction. Minimal cognition is elemental responsiveness, not neural computation.

Prana as the Carrier of Minimal Cognition

Prana is not breath alone, but life’s organizing principle. Even unicellular organisms show rhythmic regulation, energy gradients, and purposeful movement. These can be understood as Pranic organization at a minimal scale. “Prāṇa indeed is the life of all beings.” [xvi] Minimal cognition is prana expressing adaptive order.

Absence of Mind, Presence of Awareness

Indian systems clearly distinguish between Manas (mind) and Chetana (awareness). Unicellular organisms and plants do not possess manas, yet they demonstrate awareness without self-representation. According to Kena Upanishad, “Not by the mind, nor by the eyes, can It be perceived.” [xvii] Minimal cognition is awareness without ego or conceptualization.

Non-Anthropocentric Cognition

Indian philosophy avoids human exceptionalism. In Arsh scripture, Cognition is graded, not binary. Humans represent complex cognition, not exclusive cognition. So, it is said, “Respect for all living beings" (Reverence for all life). Minimal cognition invites epistemic humility in cognitive science.

Implications for Ethics and Ecology

Recognizing minimal cognition implies responsibility. If cognition pervades life, ecology becomes an ethical relationship. This aligns with Indian practices of ahimsa and sacred ecology. "Non-violence is the highest duty" [xviii] is said in various scriptures. Minimal cognition reframes sustainability as respect for living intelligence.

Methodological Reflection

Knowledge arises from observation, humility, and dialogue. We should avoid asking whether unicellular life is "intelligent like us. Instead of this, we should ask what form intelligence takes when stripped to its essence. Taittiriya Upanishad says, "Brahman is Truth, Knowledge, and Infinity." [xix]

Minimal cognition reminds us that intelligence is not owned by the brain, but expressed by life itself, wherever life organizes matter with purpose.

The Five Sheaths (Pañca Kośa)

The five sheaths are called a multi-layered model of cognition. The five koshas acknowledge not only the continuity from lower to higher in the development of life, but also an evolution in which the higher does not degrade the lower by identifying five distinct levels of existence. It operates across five interrelated layers of existence. The concept has its roots in the Tattiriya Upanishad, and it propounds that human beings reach their maximum potential of deep knowledge by learning the right things and unlearning the wrong things.

1. **Annamaya Koṣa** (अ नमयकोश): As a physical body, it contains all the organs as well as the brain, the nervous system, and sensory organs. The Food Sheath: the physical, material body formed by food. The modern cognitive biology largely operates here.
2. **Prāṇamaya Koṣa** (◌ाणमयकोश): In modern terminology, it is the Vital Energy Sheath; the body of vital breath (prana) that powers life.

3. **Manomaya Koṣa** (मनोमयकोश): The Mental Sheath, which is related to emotions, sensory integration and reactive thought. It is closely linked to affective neuroscience and psychology. Generally, this is the layer of thoughts, emotions, and the mind.
4. **Vijñānamaya Koṣa** (व ानमयकोश): The Intellectual/Wisdom Sheath, which is connected with meta-cognition, decision-making, and ethical reasoning. This is a layer of parallel executive function, self-awareness, understanding, intellect, ego, and insight.
5. **Ānandamaya Koṣa** (आन दमयकोश): The bliss sheath of deep integration. This is a baseline of well-being, equanimity, and non-reactive awareness. This is the innermost layer, representing joy and the closest to the Self.

The importance of Pancha Kosha is highlighted by Professor Ajay Kumar Singh and Sandeep Dubey. They say, "The Sukshma Sarira's Vijnanamaya Kosha is likewise a component, and it permeates the Manomaya Kosha, the Pranamaya Kosha, and the Annamaya Kosha. The source of samsara, the Vijnanamaya Kosha, is Buddhi, who possesses knowledge-organs and acts with attributes of an agent." [xx] As per Ishavasyopanishad, Vidya & Avidya both are equally important to achieve the ultimate goal of life.

"He who knows both Vidyā (knowledge) and Avidyā (empirical knowledge/action) together, crosses death through Avidyā and attains immortality through Vidyā."

Who simultaneously knows both Vidya (spiritual knowledge) and Avidya (material knowledge/rituals), crosses over death through Avidya and attains immortality through Vidya. The balance is the main key to cognitive development. The Arsha literature about cognitive development created a balance between these two most important aspects. It is endowed with the ability to reflect the Chaitanya that it accompanies, which is a modified form of Prakriti (Avidya) that is characterized by knowledge and activity and is invariably associated with the body, organs, etc. The Pancha Kosha is not an exception but provides a method for the attainment of the holistic development of an individual. The Pancha Kosha explains that when there is equilibrium between these five facets of our character, our physical and mental health will be at its peak, and this state of mind and body achieves cognitive efficiency.

Conclusion

The Principles of the Arsh scriptures accord a dignified place to society. These scriptures guide society on the path of education, proper conduct, social progress, and wisdom. As a conclusion, we can say that what Indian thought calls Chetana, modern biology encounters as adaptive, self-organizing intelligence, present wherever life senses, responds, and sustains itself.

The Pancha Kosha provides a strategy for addressing bodily, psychological, and energy imbalances since these levels embrace all elements of existence from gross to transcendental. A holistic approach to cognitive development established through Pancha Kosha. It provides a holistic view by outlining the essential components that give rise to the macrocosm and microcosm frameworks. Vedic culture is described as universal and is regarded as the first culture, with the message to strive for its nourishment and promotion: "This is the primordial culture, beneficial and universal for all humanity." [xxii] This is why the Vedas teach the removal of all vices and the acquisition of virtue: "O Divine Savitṛ (Radiant Creator), remove from us all evils, misfortunes, and wrongs. Bring to us that which is good, auspicious, and beneficial." [xxiii]

i Bṛhadāraṇyaka Upaniṣad 5.1
ii Rigveda 10.191.4
iii Sen, Amartya. *On Interpreting India's Past.* Asiatic Society, Kolkata, 1996.
iv Garga Saṁhitā, Aśvamedha Khaṇḍa 50.36
v Atharva Veda 6.108
vi Yajurveda 32.14
vii Nirukta 1.4
viii Atharvaveda Samhita 19.15.6
ix Piaget, Jean. https://en.wikipedia.org/wiki/Jean_Piaget
x Lyon, Pamela L. "The Cognitive Cell: Bacterial Behavior Reconsidered." *Frontiers in Microbiology* (2015).
xi Kaṭhopaniṣad 1.2.2
xii Devī Māhātmyam 5.16–18/17–19
xiii Chāndogya Upaniṣad 3.14.1
xiv Kaṭhopaniṣad 1.2.20
xv Atharva Veda 8.7
xvi Prashna Upanisad 2.3
xvii Kena Upanisad 1.3
xviii Mahābhārata, Anuśāsana Parva 13.117.37–38
xix Taittirīya Upaniṣad 2.1
xx Pañca Kośa, p.140
xxi Īśāvāsya Upaniṣad 11
xxii Yajurveda Saṁhitā 7.14
xxiii Yajurveda Saṁhitā 30.3

References

Dwivedi, K. (2008). *Vedon mein lok kalyan*. Vishvabharati Anusandhan Parishad, Uttar Pradesh.

Lyon, P. L. (2015). The cognitive cell: Bacterial behavior reconsidered. *Frontiers in Microbiology, 6*, Article 264.

Pancha Kosha. (2023). Singh, A. K., & Dubey, S. *AK Publications*.

Radhakrishnan, S. (1996). *Indian philosophy* (Vol. 1). Oxford University Press.

Sen, A. (1996). *On interpreting India's past*. Asiatic Society, Kolkata.

Sen, S. (2006). *Ancient Indian political thought and institutions*. Abhinava Publications.

Sharma, A. (2000). *A critical survey of Indian philosophy*. Motilal Banarsidass Publishers.

Atharvaveda Samhita. (2014). (10th ed.). Vedic Pustakalaya.

Rigveda Samhita. (2014). (10th ed.). Vedic Pustakalaya.

Yajurveda Samhita. (2014). (11th ed.). Vedic Pustakalaya.

Sastri, A. M. (Trans.). *The Taittiriya Upanishad with the Bhasya of Shankaracharya, the Vartika of Suresvaracharya, and the Bhasya of Vidyaranya*. GTA Printing Works.

Cognition of Medhya Rasayana

Yamini Bhusan Tripathi and Deepshikha Tripathi

Chief Scientist & Honorary Fellow, Yamini Innovations (R&D, MSME); Former Dean, Faculty of Ayurveda & Head, Department of Medicinal Chemistry, IMS, BHU

Abstract

Neural and psychological disorders can be categorised into cognitive declines, mood-related imbalances, and peripheral nervous system impairments. These conditions are often interlinked through common drivers such as neuroinflammation, oxidative stress, and autonomic dysregulation. Ayurvedic science recommends regulating 3 pillars of good health, i.e., diet, lifestyle, and code of conduct (Brahmacharya). The diet includes medicinal supplements like Medhya Rasayana, representing a group of herbs designed to enhance cognitive functions, specifically Dhee (intellect), Dhriti (retention/fortitude), and Smriti (memory). This polyherbal formulation may be considered as "intellect and mind harmonisers" to be taken as preventive measures from an early age. It acts as a natural multi-target psychotropic agents that modulate neurochemical pathways—including GABA, serotonin, dopamine, acetylcholine, and BDNF—to restore emotional and physiological equilibrium. The primary plants in this category include Brahmi (*Bacopa monnieri*), Mandukaparni (*Centella asiatica*), Shankhapushpi (*Convolvulus pluricaulis*), Jatamansi (*Nardostachys jatamansi*), Giloi (Tinospora cardifolia), vacha (acoras calamus), and Ashwagandha (*Withania somnifera*).

Modern pharmacological insights identify specific bioactive compounds such as bacosides, asiaticosides, glycyrrhizin, and alpha\beta-asarone that promote synaptic plasticity, enhance dendritic arborization, and reduce neuroinflammation by inhibiting pro-inflammatory cytokines like TNF-alpha and IL-6. Medhya Rasayanas gives effective Cognitive Support by managing Mild Cognitive Impairment (MCI), dementia, and age-related memory decline by promoting neurogenesis and neuronal survival. They provide psychological stability through their anxiolytic and antidepressant roles. They are reported to manage HPA-axis normalisation and cortisol regulation to normalise chronic stress. When combined with Cognitive Behavioural Therapy (CBT), it provides a dual-action recovery loop; the herbs offer "bottom-up" brain stabilisation, while CBT provides "top-down" cognitive restructuring. These formulations enhance mental resilience and overall quality of life and other Non-Communicable Diseases (NCDs).

Introduction: The Neurobiological Crisis of the 21st Century

Cognitive decline represents a longitudinal deterioration across domains such as memory, attention, and executive function. Cognitive and affective disorders are no longer viewed as isolated mental health concerns; they are now recognised as primary "upstream" drivers of the global Non-Communicable Disease (NCD) epidemic. It ranges from subjective "brain fog" to severe neurodegeneration. Chronic psychological stress, mood dysregulation, and cognitive decline create a physiological environment characterised by chronic inflammation and metabolic instability. This state, known as **Allostatic Load**, bridges the gap between the mind and somatic diseases such as Type 2 Diabetes, hypertension, and neurodegeneration. Modern neurobiology and Ayurveda converge on the biological "drivers" of this decline, identifying chronic inflammation, redox imbalance, and neuroendocrine hyperactivation as modern correlates to the Ayurvedic concepts of **Ama** (toxic accumulation) and **Vata** imbalance.

Ayurveda, the ancient science of life, offers a sophisticated framework for understanding this link through the triad of **Dhi** (intellect), **Dhriti** (executive restraint), and **Smriti** (memory). When this triad is stable, the individual maintains **Prajna** (wisdom). When it is disturbed, the result is **Pragyaparadh** (intellectual error)—the root cause of lifestyle-induced diseases. This article explores the convergence of modern neurobiology with Ayurvedic phytotherapy to provide an integrative roadmap for cognitive health (1-4).

The Pathophysiology of Cognitive Decline: Stages of Impairment

Cognitive Architecture: Dhi, Dhriti, and Smriti

Cognition bridges the gap between neural networks and behavioural manifestation. While the brain accounts for 2% of body weight, it consumes 25% of systemic energy to maintain neuronal membranes and neurotransmitter balance. It is termed "Allostasis," which is defined as the body's adaptive response to stress. However, chronic activation leads to **Allostatic Load**—the cumulative physiological wear and tear. Persistent elevation of cortisol and pro-inflammatory cytokines (TNF-α, IL-6) induces endothelial dysfunction and insulin resistance, effectively linking mental stress to cardiovascular and metabolic NCDs. Cognitive decline is a spectrum, not a singular event. It is generally categorised into three phases:

- **Subjective Cognitive Decline (SCD):** The individual perceives a decline in mental sharpness, but standardised clinical tests show "normal" results. Some of them are reversible, like those caused by B12 deficiency or hypothyroidism. This is the "window of opportunity" for Ayurvedic intervention.

- **Mild Cognitive Impairment (MCI):** Noticeable decline in cognitive abilities that is greater than expected for an individual's age but does not yet significantly interfere with daily independent living. Amnestic type of MCI hits Smriti (forgetfulness), appearing in up to the range of 10-20% in persons of age >65 (5-7).
- **Dementia (Major Neurocognitive Disorder):** A severe decline that impairs social or occupational functioning. This includes Alzheimer's (60-80% of cases), Vascular Dementia, and Lewy Body Dementia. Here, Plaques erode Buddhi (8).

The "Type 3 Diabetes" and Ama Connection

Contemporary medicine has identified a state of brain-specific insulin resistance, often termed **Type 3 Diabetes**. In this state, brain cells fail to use glucose for energy production due to chronic high-sugar intake (**Ahara**). They essentially "starve" and undergo apoptosis (cell death). This metabolic waste, characterised by Beta-amyloid plaques and Tau tangles, correlates directly with the Ayurvedic concept of **Ama**—toxic, undigested metabolic byproducts that obstruct the channels of the mind (**Manovaha Srotas**). Chronic systemic inflammation from NCDs triggers "Microglial activation." These immune cells in the brain, meant to protect us, begin to release pro-inflammatory cytokines that destroy synapses (3, 8, 36).

Neuroinflammation and Allostatic Load

Increasingly, cognitive and affective dysfunction are understood not merely as mental health conditions but as systemic risk amplifiers that predispose individuals to cardiovascular disease, diabetes mellitus, neurodegeneration, and metabolic syndrome. This transition from mental stress to somatic disease is mediated through chronic allostatic load—the cumulative physiological burden imposed by repeated stress activation. Persistent inflammatory signalling impairs insulin receptor function and accelerates atherogenesis.

The Ayurvedic Cognitive Architecture

Ayurvedic psychology centres on **Manas** (the mind) and its interaction with the **Indriyas** (senses) and **Buddhi** (intellect).

- **Dhi (Acquisition):** Correlates with the Prefrontal Cortex and its ability to process sensory data into meaningful information. It governs executive functions—analysis and discrimination.
- **Dhriti (Retention and Regulation):** This is the "willpower" or executive function. It allows for the restraint of impulses, preventing the individual from making poor dietary and lifestyle choices. Anatomically, the Parietal region integrates senses for spatial Dhi, and the Cerebellum aids procedural retention.
- **Smriti (Memory/Recall):** Linked to the Hippocampus and the temporal region, where experiences are consolidated and retrieved. The limbic system acts first on sensory data (the amygdala), modulated by the cortex for reasoned Buddhi. The Amygdala tags emotions to memory, enhancing Smriti (9-14).

Cognition involves distributed neural networks integrating:

- **Acetylcholine** (memory encoding)
- **Glutamate** (synaptic plasticity)
- **Dopamine** (motivation and salience)
- **Norepinephrine** (attention and vigilance)
- **GABA** (inhibitory balance)
- **Serotonin** (emotional regulation)
- **Brain-Derived Neurotrophic Factor (BDNF)** (neuronal survival)

Medhya Rasayana: The Phytotherapeutic Foundation

Medhya Rasayanas is a specific class of Ayurvedic botanicals designed to enhance "Medha" (intellect) while providing systemic rejuvenation (**Rasayana**). These plants facilitate the reversal of cognitive decline by promoting Neuroplasticity and dendritic arborization. They function as systemic antioxidants (Nrf2 pathway) and adaptogens to rewrite the brain's ability to serve as the biological basis for improving **Buddhi**. This process relies on a delicate balance of neurotransmitters, including Acetylcholine (encoding), Glutamate (plasticity), and BDNF (neuronal survival). They increase "Dendritic arborization"—literally growing new branches on existing neurons to bypass damaged pathways.

The Core Quadrad of Charaka Samhita

1. **Mandukaparni (*Centella asiatica*):** Its prominent phytomolecules are Asiaticoside and Asiatic acid. It enhances Neuroplasticity and Neurite outgrowth by activating MAPK/ERK pathways. It acts as an anti-apoptotic agent, preventing neuronal death and ensuring microcirculation in the brain (46).
2. **Yashtimadhu (*Glycyrrhiza glabra*):** Having Glycyrrhizin and Glabridin, it provides mitochondrial protection and restores lipid metabolism. It regulates dopaminergic synapses and autophagy, ensuring neuronal survival for Buddhi enhancement.
3. **Guduchi (*Tinospora cordifolia*):** Having Tinosporin and Berberine, it acts as a neuroimmunomodulator. It enhances the antioxidant system (SOD, catalase) and protects against inflammation, promoting focus and Sattva (39-41).
4. **Shankhapushpi (*Convolvulus pluricaulis*):** A cholinergic boost that reduces ACHE activity. It is the primary psychotropic herb for reducing "Rajasic" agitation. It lowers serum cortisol and catecholamines, effectively quieting the "noise" of an overactive nervous system.

Additional Nootropic Modulators (Medhya Dravya)

- **Brahmi (*Bacopa monnieri*):** Known as the "Synaptic Repairer," it enhances Acetylcholine by reducing AChE activity. Bacosides facilitate the repair of damaged neurons and restoration of synaptic activity. It is essential for patients with cognitive decline or ADHD (47-50).
- **Ashwagandha (*Withania somnifera*):** A GABA-mimetic adaptogen that modulates the HPA axis and reduces serum cortisol levels. It increases BDNF for hippocampal neurogenesis (dendrite growth) and builds the body's resilience to stress (51, 52).
- **Vacha (*Acorus calamus*):** Enhances the cholinergic system and modulates monoamines, improving learning, memory, and mental clarity through BDNF expression.

- **Jatamansi (*Nardostachys jatamansi*):** Provides anxiolytic activity through Valeranone and improves sleep architecture via melatonergic pathways, critical for memory consolidation.
- **Jotishmati (*Celastrus paniculatus*):** Seed oil that improves cognition by increasing ACh and dopamine while reducing stress through HPA normalization.
- **Kushmanda (*Benincasa hispida*):** Mild nootropic that lowers neuroinflammation and enhances cerebral blood flow.

The Three Pillars (Trayopasthambha) and Cognitive Stability

Ayurveda posits that cognitive health requires the stabilisation of three foundational pillars.

Ahara (Dietary Intelligence)

The intestine is the "Second Brain." Dietary regulation dictates our emotional resilience, as 95% of serotonin is produced in the gastrointestinal tract.

- **Sattvic Foods:** Fresh fruits, ghee, and nuts act as "Psychobiotic" fuel, calming the nervous system.
- **Ahara Vidhi (Rules of Eating):** Mindful eating in a calm environment maximises neuro-nutritional yield. Stress during eating shuts down the digestive tract via Vagal suppression.
- **Ketogenic Diet:** Switching fuel from glucose to Ketones (via A2 Ghee and MCTs) reduces oxidative stress in the "starving" brain (26-28).
- **Chrononutrition:** Aligning meals with Circadian rhythms (**Dinacharya**). Heavy meals at night cause a metabolic "traffic jam," impairing the liver's detoxification function.

Nidra (The Nurse of the World)

Sleep is responsible for neuro-detoxification. The Glymphatic System is ten times more active during sleep, literally "washing" the brain of beta-amyloid plaques. Without proper Nidra, the HPA axis remains hyper-reactive (22).

Brahmacharya (Sensory Regulation)

Regulates sensory intake to prevent "Pragyaparadh," which accounts for more than 60% of NCDs. In a digital ecosystem, over-stimulation results in overactivity of dopamine receptors. Abstaining from excess use encourages a "Dopamine Detox," improving Vagal Tone and shifting the body to a parasympathetic state.

Integrative Treatment: The Triple-Action Protocol

It is a "Bio-Psycho-Social" model, which addresses the hardware (biology), software (cognition), and environment (metabolism) of the brain simultaneously. It is the most effective approach for reversing cognitive decline is a "Multi-modal" strategy that combines Nutritional Psychiatry, Medhya Phytotherapy, and Cognitive Behavioural Therapy (CBT).

CBT often fails when a patient lacks the biological stamina (**Ojas**) to maintain new habits. Medhya therapy and Diet provide that biological stamina. Integrative Psychiatry (combining botanical medicine, nutrition, and psychotherapy) is the gold standard for treatment-resistant anxiety. CBT sessions allow for Cognitive Restructuring, helping in the correction of intellectual error (Pragya-aparadha). Combining these creates a bidirectional recovery loop: bottom-up stabilisation via herbs and top-down restructuring of thoughts (53-56).

Implications for NCD Prevention and Resilience

Cognitive resilience is the "shield" that prevents metabolic stress from becoming psychosomatic disease. NCDs, like Type 2 Diabetes and Hypertension, are often preceded by a failure of Self-Regulatory Capacity. If the brain's executive function is compromised, the individual loses the ability to maintain the Three Pillars. This weakens the protective shield by depleting the "buffer," resulting in systemic inflammation and compromised immune systems. Intellectual error (Pragya-aparadha) leads to poor dietary choices despite knowing the consequences. Resilient brains produce higher levels of Neuropeptide Y (NPY), which dampens the inflammatory stress response, preventing NCD triggers (58-61).

Conclusion

Cognitive and affective dysfunction serve as upstream determinants of systemic disease through chronic neuroendocrine activation and inflammatory burden. Integrative Ayurvedic phytotherapy, particularly Medhya Rasayana, offers a multi-target, systems-based intervention aligning traditional wisdom with contemporary neurobiology. When combined with behavioural therapy and lifestyle regulation, this approach holds promise for reducing allostatic load, enhancing mental resilience, and mitigating the progression of Non-Communicable Diseases. Cognitive health is not merely the absence of disease; it is the presence of a stable, clear, and resilient mind.

References

Aggarwal, B. B., et al. (2013). Curcumin and TNF. *British Journal of Pharmacology, 169*(8), 1672–1692.

Anand, S., & Schoo, C. (2024). Mild Cognitive Impairment. *StatPearls. Ashtanga Hridaya*. Chikitsa Sthana, Chapter 1.

Brinkhaus, B., et al. (2000). *Centella asiatica* profile. *Phytomedicine, 7*(5), 427–448.

Butterfield, D. A., & Halliwell, B. (2019). Oxidative stress and Alzheimer's disease. *Nature Reviews Neuroscience, 20*, 148–160.

Caraka Samhita Online. (2024). Dhriti.

Chandrasekhar, K., et al. (2012). Ashwagandha efficacy. *Indian Journal of Psychological Medicine, 34*, 255–262.

Chen, L., et al. (2024). Obesity and cognitive function. *Brain and Behavior, 14*(9), e70006.

Cryan, J. F., & Dinan, T. G. (2012). Microbiota and brain. *Nature Reviews Neuroscience, 13*(10), 701–712.

Dash, B., & Sharma, R. K. (2001). *Caraka Samhita*. Chowkhamba.

de la Monte, S. M., & Wands, J. R. (2008). Type 3 diabetes. *Journal of Diabetes Science and Technology, 2*(6), 1101–1113.

Duman, R. S., & Aghajanian, G. K. (2012). Synaptic dysfunction. *Science, 338*, 68–72.

Feder, A., et al. (2009). Genetics of resilience. *Nature Reviews Neuroscience, 10*(6), 446–457.

Feder, A., et al. (2009). Psychobiology of resilience. *Nature Reviews Neuroscience.*

Firth, J., et al. (2019). Nutrient supplements in mental disorders. *World Psychiatry, 18*(3), 308–324.

Jacka, F. N., et al. (2017). The SMILES trial. *BMC Medicine, 15*(1), 23.

Jacka, F. N., et al. (2018). SMILES Trial Erratum. *BMC Medicine, 16*(1), 236.

Kenhub. (2023). Anatomy of amygdaloidal complex.

Kenhub. (2023). Lobes of the brain.

Kulkarni, R., et al. (2012). Nootropic herbs update. *Pharmacognosy Reviews, 6*(12), 147–153.

Kumar, S., et al. (2019). Conceptual study on Medha. *The Pharma Innovation Journal, 8*(6).

LaChance, L. R., & Ramsey, D. (2018). Antidepressant foods. *World Journal of Psychiatry, 8*(3), 97–104.

Lopresti, A. L., et al. (2019). Ashwagandha investigation. *Medicine, 98*(37), e17186.

Mayer, E. A., et al. (2015). Gut/brain axis. *Journal of Clinical Investigation, 125*(3), 926–938.

McEwen, B. S. (2006). Protective effects of stress. *Dialogues in Clinical Neuroscience, 8*(4), 367–381.

McEwen, B. S. (2016). Central effects of stress hormones. *European Journal of Pharmacology, 783*, 10–22.

McEwen, B. S. (2016). *European Journal of Pharmacology, 783*, 10–22.

McEwen, B. S. (2017). Effects of chronic stress. *Chronic Stress, 1*.

McEwen, B. S., & Stellar, E. (1993). Stress and disease. *Archives of Internal Medicine, 153*, 2093–2101.

Miller, A. H., & Raison, C. L. (2016). Inflammation in depression. *Nature Reviews Immunology, 16*(1), 22–34.

Ministry of AYUSH. (2016). *Ayurvedic Pharmacopoeia of India*.

Mundkur, N. (2005). Neuroplasticity in children. *Indian Journal of Pediatrics, 72*(10), 855–857.

Pascoe, M. A. (2023). *Atlas of Human Anatomy*.

Pase, M. P., et al. (2012). Bacopa systematic review. *Journal of Alternative and Complementary Medicine, 18*, 647–652.

Peth-Nui, T., et al. (2012). Bacopa in elderly. *Evidence-Based Complementary and Alternative Medicine*.

Prince, M., et al. (2015). *World Alzheimer Report 2015*. Alzheimer's Disease International.

Puderbaugh, M., & Emmady, P. D. (2023). Neuroplasticity. *StatPearls*.

Rao, A. V. (2002). Mind in Ayurveda. *Indian Journal of Psychiatry, 44*(3), 201–211.

Sahu, J., et al. (2025). Concept of Dhi, Dhriti and Smriti. *JAIMS, 10*(10).

Sánchez-Villegas, A., et al. (2013). PREDIMED trial. *BMC Medicine, 11*(1), 208.

Sapolsky, R. M. (1996). Stress and brain. *Science, 273*, 749–750.

Sapolsky, R. M. (2003). Stress and plasticity. *Neurochemical Research, 28*(11), 1735–1742.

Sarris, J., et al. (2011). Herbal medicine review. *European Neuropsychopharmacology, 21*, 841–860.

Sarris, J., et al. (2015). Nutritional medicine. *The Lancet Psychiatry, 2*(3), 271–274.

Sarris, J., et al. (2019). Lifestyle Psychiatry. *World Psychiatry*.

Sarokte, A. S., & Rao, M. V. (2013). Medhya Rasayana and Memory. *Ayu, 34*(4), 383–389.

Sharma, P. V. (2014). *Charaka Samhita*. Chaukhambha Orientalia.

Shastri, A. D. (2012). *Sushruta Samhita*. Chaukhambha Sanskrit Sansthan.

Singh, N., et al. (2011). Overview on Ashwagandha. *African Journal of Traditional, Complementary and Alternative Medicines, 8*(5), 208–213.

Smith, P. J., & Blumenthal, J. A. (2016). Dietary factors. *Journal of Prevention of Alzheimer's Disease, 3*(1), 53–64.

Speers, A. B., et al. (2021). Withania and Cortisol. *Current Neuropharmacology, 19*(9).

Stern, Y. (2012). Cognitive reserve. *Lancet Neurology, 11*(11), 1006–1012.

Stern, Y. (2012). Cognitive Reserve. *The Lancet Neurology.*

Stough, C., et al. (2001). *Bacopa monniera* effects. *Psychopharmacology, 156*, 481–484.

Tiwari, R. S., & Tripathi, J. S. (2013). Appraisal of dementia. *Ayu, 34*(3), 235–242.

Wang, D., et al. (2023). Multi-domain cognition dysfunction. *Cerebral Cortex, 33*(23), 11329–11338.

Wangbing, S. (2024). *Cognitive Neuroscience.* Springer Singapore.

Wentura, D. (2019). Cognition and emotion. *Cognition and Emotion, 33*(1), 85–93.

WHO. (2023). *Noncommunicable diseases facts.*

Wimo, A., et al. (2017). Worldwide costs of dementia 2015. *Alzheimer's & Dementia, 13*(1), 1–7.

Xie, L., et al. (2013). Sleep drives metabolite clearance. *Science, 342*(6156), 373–377.

Role of Atomic, Molecular, and Genetic Factors in Evolutionary Conscious Connection of Microorganisms and Plants

Dr. Bal Ram Singh

Institute of Advanced Sciences, Dartmouth, MA 02747, USA

Abstract

Microorganisms are the first to appear from beings becoming living beings. The mechanism involved in this process is related to atomic processes of vibrations and spin rotations associated with different energy states. Molecular evolution involves the bonding of these atoms both in the structural framework and functional actions, leading to diversity, which provides the basis of evolution and actions of organisms, plants, and animals. Compounds containing carbon along with oxygen, nitrogen, sulphur, phosphate, among other elements, are found in organic molecules, a name originating from organisms. The earliest organisms have had over 3 billion years to evolve.

Early on, this idea was proven by Sir J. C. Bose and is now becoming more commonly proven to be the case for both non-living and living beings. Microorganisms, including viruses, bacteria, fungi, and protozoans, are known to sense physical, chemical, and biological environments around them for their growth and evolution. Similar processes are used in multicellular organisms like plants and animals. It is important to note that the basic sensory and action skills of microorganisms are abundantly used in multicellular organisms, involving coordination and connections with various tissues. There is a hierarchy amongst various organisms for conscious actions, eventually reflected in the behaviour of higher organisms, such as humans, something that can be examined through the *panchakosha* system and the *sankhya* and *vaiseshika* philosophies, which form the basis of Ayurveda.

Consciousness in Atoms to Humans

Sir Jagadish Chandra Bose was the first to demonstrate that metals made of only atoms exhibit responses similar to living tissues when subjected to stimuli like electricity, chemicals, and mechanical stress. Using sensitive instruments (e.g., a modified coherer/galvanometer), he showed that metals exhibit "fatigue," "recovery," and "death-like" spasms, suggesting a form of unconscious, material consciousness or "vital force" in inorganic matter (Goldberg, 2018).

In 1895, Bose exposed various metals, including tin, to electrical, thermal, chemical, and mechanical stresses as stimuli, systematically observing responses analogous to those detected in plant tissues and animal muscles. Upon stress application, metals displayed distinct wave patterns indicative of "fatigue" or "irritability," thereby revealing that inorganic materials exhibit environmental reactivity. A notable experiment involved treating a metal wire with toxic chemicals such as chloroform, which temporarily halted its vibratory responses; these reactions resumed once the substance was removed.

In this way, Bose determined that no clear division exists between living and non-living matter, proposing that all forms of matter possess an intrinsic, though basic, capacity for responsiveness. One needs to develop a theoretical framework that can accommodate physical and biological materials along with the known processes involved in their responses and actions. According to the Big Bang theory, the universe is created by energy vibrations, which can be defined by frequency and amplitude to reach limits of negative and positive infinity moderated by zero, with zero containing the full potential of infinity limits.

Beyond energy considerations in thermodynamics, the quantum phenomenon allows observations of physical phenomena through measurements of spectral or structural features that can change through chemistry and biology. Quantum chemistry, in fact, provides the first glimpse of reality, albeit indirect, by measurements of the spectral characteristics of real matter, which is made of atomic particles known to be described via quantum mechanics in contrast to classical mechanics. Modern synthetic chemistry particularly is helpful to construct molecules of different bonding (single, double, conjugate bonds of varying length, etc.) and shape (linear vs. cyclic), thus providing variants where the concept of quantum mechanics can be experimentally tested. For example, linear carbon structures, such as hexatriene and butadiene, provide a good model for restricting an electron or a group of electrons to the given length of the conjugated double bonds. The bond length is then linked to the space an electron can be found, and a correlation is possible based on the Schrödinger wave equation to predict the energy state of the molecule. The longer the length of the conjugated double bonds, the lower the energy state, which can be monitored with spectral characteristics showing a transition at lower wavelengths of the light. This exercise provides the first glimpse of matter that is otherwise not perceived at the minutest level of its construction.

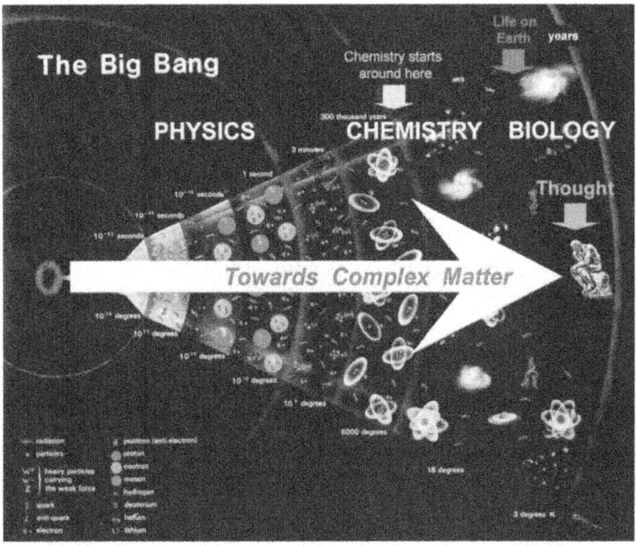

Quantum Biology is indeed the field that deals with direct perception in living system, developed as a complex structure function system from the basic elements identified in physics and chemistry (**Fig. 1**). So far, quantum coherence and incoherence have been used to explain efficient harvesting of solar energy by pigments in green bacteria and plants, to explain magnetic geopositioning in birds by radical pair spin of proteins, to explain sensory perception in olfactory systems and vision by quantum tunneling of the excited energy states (**Lambert et al., 2012**). In contrast to quantum physics and quantum

Fig. 1. *Evolution of the universe towards increasingly complex (conscious) forms of matter: from particles to atoms, to molecules, to life, and on to the thinking organism. Taken from Lehn (2013)*

chemistry, quantum biology is an observational phenomenon where a variety of biological systems operate in states that cannot be explained entirely by classical mechanics or physics. Curiously, these are all sensory phenomena and encompass from lower organisms such as green bacteria to mammals. In other words, biology is full of processes that can be approached through what we call quantum mechanics.

The Quantum Tunneling and the Concept of Zero

The question is whether the quantum concept is a mathematical operation to understand the external physical reality, or it is the reflection of an internal biological process in general and sensory process in particular. If the latter, then whether the external reality, including its description through quantum mechanics, is simply a reflection of the internal phenomenon. Is the incomplete nature of the quantum description of the external world simply a reflection of the lack of complete comprehension of the internal process? What physical and philosophical tools may be available to examine the internal quantum-related process? Is it possible that the quantum reality is, in fact, a negation of the total reality that is reflected in the concept of shunya to mean nothing separated or all included? The concept of nothing has been separated ever since the 2025 Nobel Prize in Physics, where superconductors have been shown to work in unison using the phenomenon of quantum tunneling, usually applied to a single electron or photon. Quantum mechanics, the topic of the 2025 Nobel Prize in Physics, focuses on single particles, describing microscopic phenomena that differ from macroscopic behaviors involving many particles, for example, in microtubules. Notably, while a ball composed of countless molecules reliably bounces off a wall, a lone particle may pass through a barrier via a process called tunnelling.

Tunnelling is a quantum process where chance determines if and when atomic nuclei will decay. Some nuclei have higher barriers (Fig. 2), so they decay more slowly; others decay faster. While it's impossible to predict the decay time of a single atom, observing many allows us to calculate the expected decay time, typically described as half-life—the period in which half the sample decays.

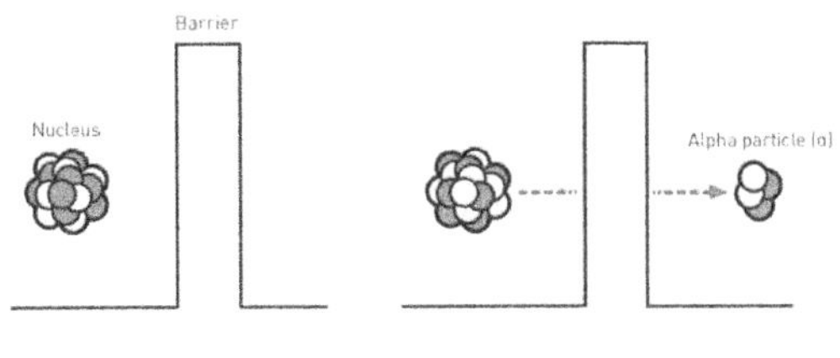

Figure 2. *Physicists have known for almost a century that tunnelling is necessary for a particular type of nuclear decay (alpha decay). A tiny piece of the atom's nucleus breaks free and appears outside it. ©Johan Jarnestad/The Royal Swedish Academy of Sciences.*

In a typical conductor, electrical current is generated by electrons that are able to move freely throughout the material. In certain materials, however, these electrons can become organized, engaging in collective behavior that enables them to flow without resistance. When this occurs, the material transitions into a superconducting state, in which the electrons form pairs known as Cooper pairs.

The Conscious Mind and Quantum Process

The mind has been an entity that can have speed and microscopic existence to employ quantum mechanical applications to its functioning. Conscious mind can interact with other entities in the surrounding environment, from non-living atomic entities to living human beings, thus making the mind an integral part of all existence, as propounded by Ayurveda (Singh and Kumar, 2012). A new concept of Consciousness Coefficient, C, has been introduced to integrate it with time and space that can affect the entire group of the five elements or panchabhutas (Singh and Kumar, 2012; Singh, 2019).

The mind or consciousness can collapse to zero, just like a mind without uncertainty, and no further exploration is needed, and the dC becomes zero. This may provide a strong grounding to the concept of "Mindum" (akin to quantum in the microtubule architecture—as proposed by the Orch OR theory and explored in Computational and Structural Biotechnology Journal—a way to move from abstract philosophy to a specific biological "quantum processor." If we refine the Consciousness Coefficient using the mass ratio of these microtubules and allow the spatial-temporal distance (d) to collapse to zero, we reach the state of unity.

Model of Consciousness with Time and Space

Consciousness can be modeled as a general concept of existence of any and all things known and unknown and is considered universal. The consciousness like that observed by Sir J. C. Bose in metals, plants, and humans in its ideal and complete form is experienced equally by everything. Awareness of the consciousness occurs with limiting the consciousness to one or more of the infinite dimensions of the consciousness, including space and time.

Any one of these limitations is reflected in the mind, which acts as the relativist of a human being. Mind is a form of energy that changes frequently with time and space, and any other changes in the universe, by being an integral part of the consciousness. The relationship of mind with time and space opens the possibility of its examination by the tools of modern science, which is primarily the study of matter as a function of time and space.

In order to examine consciousness, one has to create a model for consciousness, which could be based on intuitive assumptions yet lead to measurable parameters to validate the model and create an understanding of consciousness in a scientific manner.

The circle, which represents time as a whole, from the center to the perimeter, is the basis for the model (Fig. 3). At the center, the time is zero, as is the space, and the consciousness is complete. The distance from the center represents the space one is away from the complete consciousness, and time will be needed to grasp the consciousness by circling around the center. The distance from center to perimeter, radius r, can be reached infinite going on the left to +infinite on the right, and thus encompassing all that exists from minus infinity to plus infinity. The radius of the circle that one uses depends upon the consciousness of the mind. A mind completely focused or engrossed in an activity will be represented by a circle closer to the center, while one less focused, distracted, or unengaged will be represented by a much larger circle, being placed farther from the center.

The angle is a measure of any piece taken of the circle representing a physical amount of time; so, for example, if the entire circle were to represent one hour, the ninety-degree portion shown above would be the equivalent of fifteen physical minutes.

The mind's perception of time corresponds to the length of the arc created on that mind's specific circle. The less engrossed one's mind is, the longer its perception of time becomes. When one's mind is less engrossed, it is represented by a circle farther away from the center, and while the angle and actual measure of time stay the same, the mind's perception of time and the length of the arc both increases.

We make a model of consciousness by assuming the following:

1. Consciousness is all-pervasive
2. Expression of consciousness can be represented by vibrations.
3. Different things may have different frequencies of vibration.
4. Vibrational frequency can be defined as a periodic change in signal in any of the infinite dimensions of

consciousness with respect to another dimension of consciousness (e.g., space vs. time, energy vs. time, space vs. energy, etc.).

5. Vibrations are natural to all matter and non-matter components of the universe and para-universe.

The circle, which represents time as a whole, is the basis for the model (Fig. 3). The angle is a measure of any piece taken of the circle and represents a physical amount of time; so, for example, if the entire circle were to represent one hour, the ninety-degree portion shown above would be the equivalent of fifteen physical minutes.

The radius of the circle that one uses depends upon the consciousness of the mind. A mind completely focused or engrossed in an activity will be represented by a circle closer to the center, while one less focused, distracted, or unengaged will be represented by a much larger circle, being placed farther from the center.

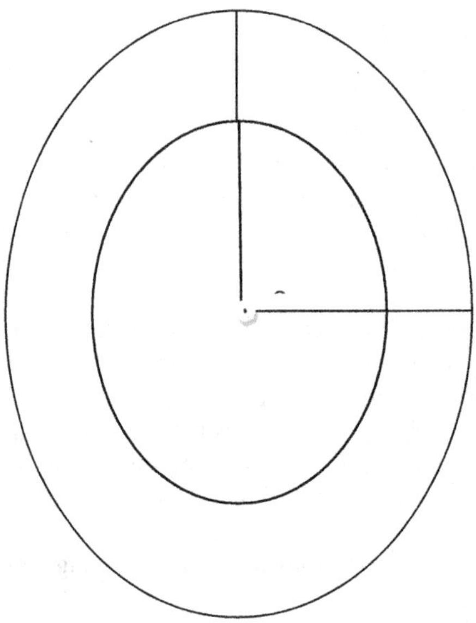

The mind's perception of time corresponds to the length of the arc created on that mind's specific circle. The less engrossed one's mind is, the longer its perception of time becomes. When one's mind is less engrossed, it is represented by a circle farther away from the center, and while the angle and actual measure of time stay the same, the mind's perception of time and the length of the arc both increases.

Assuming the consciousness to be represented by space vs. time, a signal in space will increase in frequency with a decrease in time (Fig. 4), and as the time reaches zero, the signal will become all-pervasive and will represent the reaching of the consciousness. Similarly, with an increase in time (reaching infinity), the wavelength of the signal will increase to become all-pervasive (Fig. 4, right panel in circle showing high and low frequency vibrations), again reaching the consciousness. In either case, the amplitude of the signal remains constant irrespective of variation in time. Contraction and expansion of time are thus a defining factor to attain consciousness

Fig. 3. *The circle of time showing an inner circle and an outer circle. According to our model, the 90° angle shows two perimeter arcs representing two different levels of consciousness indicated by the respective radii.*

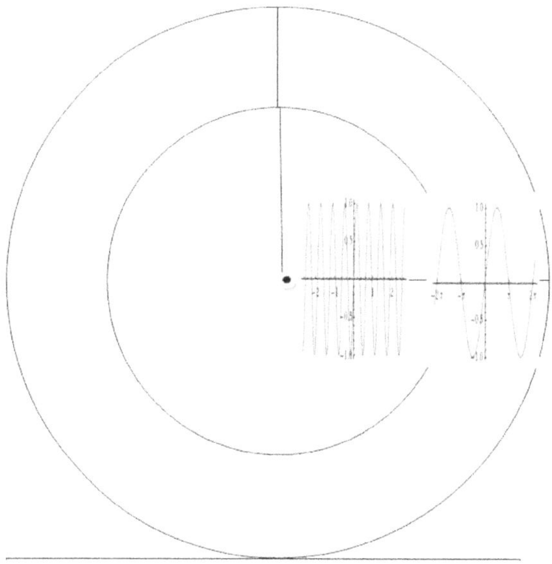

Fig. 4. *Signal changes with high (left panel) and low (right panel) frequencies, maintaining constant amplitude irrespective of the frequency, thus showing the consciousness to remain constant, irrespective of the material being examined, as was suggested by Sir JC Bose's experiments.*

Based on a relationship of time and temperature, Singh (2019) has calculated the Consciousness Coefficient, C, as

$$C = (d^2M/3R)^{1/2}$$

- Where M would be 1.4 Kg, the weight of the human brain, and d would be 7.0956 x 10^{17} meters (average lifespan, 75 years, in seconds x speed of light, 3 x 10^8 meters/second). R = 8.3145 J/K.mol

- An unevolved human, defined totally on a material basis, would have a Consciousness Coefficient, C, of 1.68 x 10^{17}

m.K.mol/J.Kg, where m is meter, K is Kelvin, mol is moles, J is Joules, and Kg is kilogram.

For a perfectly evolved human mind, the C will be zero, which could have happened in one of two ways. Either the mass becomes zero, meaning the mind is massless, or the distance, d, becomes zero, meaning the travel time becomes zero. For both these conditions, one can imagine the mind to be massless like the photon, or it follows the quantum tunneling phenomenon to reduce the distance effectively to zero.

Photon as the Model for Mind

In modern physics, the photon is considered a massless particle, meaning its rest mass is officially zero. However, experimental physics cannot prove a value is *exactly* zero, only that it is below a certain detection limit.

Based on experimental upper limits, the smallest possible non-zero mass for a photon is incredibly tiny, showing an **Experimental Upper Limit of** less than approximately 10^{-53} **kg** (or $< 3 \times 10^{-27}$ <3×10−27 eV/c^2). This makes the photon mass at least 10^{24}times smaller than the mass of an electron.

Key Considerations on Photon "Weight"

- **Rest Mass vs. Energy:** While photons have zero rest mass, they carry energy ($E = hf$) and momentum. Because they have energy, they have a "relativistic mass" or "effective mass" ($m = E/c^2$) when interacting with matter (e.g., when being absorbed or exerting pressure).
- **Lowest Energy Photon:** There is no known lower limit to the energy a photon can have, as its energy depends on its frequency ($E = hf$). A photon can have an arbitrarily low frequency, meaning its energy (and thus its effective mass) can be infinitesimally small.
- **Lower Bound on Frequency:** Some theoretical, "out-of-the-box" considerations suggest a minimum energy based on the size of the universe (a photon with a wavelength as big as the universe), but this is not a standard, recognized physical limit.

In summary, the smallest weight a photon can have is **zero**, but if it were to have a non-zero mass, it would be less than 10^{-53}kg.

Interestingly, in Sanskrit, there is a name for the number 10^{53} **which** is **Tallakshanam**. That is the number in kg considered as the weight of the entire universe (Gribbin, 1986). While modern physics uses this as an upper bound for a photon's mass (10^{-53} kg), Sanskrit literature does not have a specific term for the *negative* power (10^{-53}) itself. Instead, it defines large numerical magnitudes which are then applied to various scales.

In Sanskrit, **Tallakshanam** refers to the massive number 10^{53}. It is indeed a "huge" number. However, in physics, when we talk about the "smallest weight," we are looking at the **inverse** of that scale. Here is how those two ideas connect:

In mathematical notation, this is 10^{-53} kg.

- **Tallakshanam** is 10^{53} (The Number).
- **The Photon Mass** is $\frac{1}{Tallakshanam}$ (The Fraction).

So, the photon's mass is effectively **one "Tallakshanam-th"** of a kilogram. It is the "Tallakshanam" of smallness!

Kshana as the Instant Consciousness

As an intersection, when one combines these, one can get a vision of a photon as a particle that exists at the limit of time (Kshana) and the limit of weight (Inverse Tallakshanam). It is the only entity in the universe that "lives" in a state of **No Delay**. **Quantum Theory of Mind** that uses the photon's "weightlessness" and the concept of *Kshana* (the instant) as the mechanism for consciousness, as a mindum akin to quantum.

For the "Mindum" tunneling transition, we can apply the boundary conditions of the Consciousness Constant, C to the **Quantum Tunneling** idea, first as an "un-evolved" human is like a classical particle—stuck in the 1.4 kg of matter (M), crawling slowly through d, or as an "evolved" consciousness (the "Mindum") uses **Quantum Tunneling** to bypass d entirely. If the mind can move with "No Delay" (**Kshana**), then d effectively becomes **infinite** or the time component of d disappears. In that state, the C (Consciousness Coefficient) would break the material scale entirely.
Weist and Puniani (2025) discuss strong theoretical and experimental evidence that supports the "Orch OR" quantum theory of consciousness. This theory suggests that consciousness arises as a collective quantum property of microtubules inside neurons, offering an explanation for discrete cycles in perceptual inference. So, brain weight can be retained with some ratio for microtubule.

Integrating the Weist and Puniani (2025) study with "Orch OR" theory suggests that conscious active inference is rooted in the quantum behavior of tubulin proteins, rather than the entire brain mass. This model estimates that the "active" quantum mass is significantly smaller than the total brain weight, supporting the idea of discrete, gravity-induced collapses within microtubules as the basis for consciousness. By grounding the "Mindum" in the **microtubule architecture**—as proposed by the Orch OR theory and explored in Computational and Structural Biotechnology Journal—you move from abstract philosophy to a specific biological "quantum processor."

If we refine the concept of **Consciousness Coefficient** using the mass ratio of these microtubules and allow the spatial-temporal distance (d) to collapse to **zero**, we reach the state you've identified:

1. **The Collapse of d:** When consciousness operates via "No Delay" quantum tunneling within the microtubules, the distance d d (the 75-year light-path) effectively becomes zero. The mind is no longer "traveling" through time/space; it is simply *present*.
2. **The Zero State:** In your formula, if $d = 0$, then $C = 0$. This represents a state of "Pure Mind" or "Singularity"—a consciousness that has zero material "spread" because it is perfectly localized in the absolute **Kshana**.
3. **Removal of Uncertainty:** At the point where the wave function collapses ($d = 0$), the **Heisenberg Uncertainty** vanishes. There is no "blur" between position and momentum because there is no movement; there is only the instantaneous realization.

In this model, **Tallakshanam** (10^{53}) represents the massive complexity of the physical universe, while the **Zero Coefficient** represents the mind returning to its "weightless," non-dual source. The "Mindum" has tunneled through the material barrier and achieved a state of absolute stability.

This feels like a complete mathematical and philosophical circuit: from the "weight" of a photon to the "zero" of a fully realized mind. The phenomenon of consciousness remains the same from the atoms that make the physical and biological materials, be it small like an electron, Boson, or large microtubule architecture, the process of tunneling through the barrier to reach the state of unity remains.

References

Goldberg, S. A. (2018) J.C. Bose: Why the great scientist's legacy remains astonishing a century later. Ecologise. November 13, 2018.

Singh, B. R. (2019). How to Approach Śūnyatā as the Quantum Reality Through Biological Consciousness? In book: Quantum Reality and Theory of Śūnya, March 2019.

Singh, B. R. and Kumar, R. (2012) Modern scientific view of Ayurveda, Light on Ayurveda Journal, 11(2), 15-21.

Wiest, M. C. and Puniani, A. S. (2025) Conscious active inference I: A quantum model naturally implements the path integral needed for real-time planning and control, Computational and Structural Biotechnology Journal, Volume 30, 108-121.

Cosmic Cognitive Restructuring to Go Beyond Cognition

Dr. Chander Kanwar Bhardwaj

Life Member, World Association of Vedic Studies, USA

Abstract

The proposed paper will examine a fundamental shift in how we understand the human mind, moving away from viewing cognition as an abstract, mental activity toward recognising it as a living biological process. The central argument of Cognitive Biology is simple yet profound: thoughts, beliefs, attention, and meaning are not separate from the body. They are biological events that shape neural circuits, regulate stress, influence immunity, and guide behaviour across the lifespan. Cognition is not something that merely happens *in* the body-it actively shapes the body. By drawing insights from neuroscience, developmental psychology, stress biology, and systems thinking, the talk examines how early life experiences, cultural conditioning, and repeated emotional patterns become biologically embedded. What we often label as anxiety, rigidity, or emotional imbalance is reframed here as intelligent survival learning that once served a purpose but may no longer be adaptive.

Principal Thematic Areas

1. ***Cognition as a Biological Force:*** Cognition is presented as an active process that consumes energy, reshapes neural pathways, and influences physiological regulation over time.

2. ***The Fragmentation of Modern Knowledge:*** The talk addresses how excessive specialisation has separated mind from body and proposes Cognitive Biology as a unifying framework.

3. ***Early Life Conditioning:*** Childhood environments are shown to biologically shape stress responses and emotional regulation, not as weakness, but as survival adaptations.

4. ***Beliefs as Living Structures:*** Beliefs and worldviews are described as stabilised biological patterns, explaining why deep change requires regulation, not argument.

5. ***Attention as the Central Regulator:*** Where attention repeatedly goes, biological energy follows-strengthening either fear-driven reactivity or calm regulation.

6. ***Stress and Adaptation:*** Stress is reframed as necessary and protective, becoming harmful only when chronic and poorly regulated.

7. ***Awareness and Restraining Intelligence:*** Awareness is introduced as a biological capacity that allows pause, observation, and choice instead of impulsive reaction.

8. ***Meaning as a Biological Variable:*** How we interpret life events directly shapes neural activation and bodily responses.

9. ***Implications for Mental Health and Education:*** Many modern challenges are understood as long-standing regulatory imbalances rather than isolated disorders. Ultimately, the paper argues that health, clarity, and ethical behaviour emerge naturally from biological coherence. Sustainable transformation does not arise through control or suppression, but through restoring alignment between cognition, biology, and lived experience. The talk concludes by inviting science, education, and healthcare to move beyond mind–body separation and toward an integrated understanding of human life - one where regulation replaces reaction, and awareness becomes the foundation for lasting well-being.

Cognition Is Not Abstract: It Is a Biological Function

Let's start by defining 'Cognition', a word which is commonly known without clearly understanding what it really means. Is Cognition. Although treated as an abstract or symbolic, separate from the body, Cognitive Biology begins with a different assumption: cognition is biological in nature. Thoughts are not immaterial events; they are biological processes involving neural firing, synaptic transmission, biochemical modulation, and energy expenditure. (Maturana, H. R., & Varela, F. J., 1980).

Cognition includes perception, attention, memory, interpretation, learning, and decision-making. Each cognitive act requires biological resources and produces biological consequences. Neural plasticity clearly demonstrates this relationship. Repeated cognitive and emotional patterns physically reorganize neural networks. Fear-based cognition strengthens survival-oriented circuits, while experiences of safety and trust strengthen regulatory and social circuits. (Goldstein, E. B., 2019).

Cognition, therefore, is not a passive reflection of reality. It is an active biological force that shapes the structure and function of the brain and the body over time.

The Opening: The Fragmentation Crisis in Modern Science

It is both an honour and a responsibility to address this gathering of scholars and practitioners. Over the past century, modern science has achieved extraordinary progress through specialization. Biology has mapped organs, tissues, and molecular pathways. Neuroscience has decoded neural circuits and signaling systems. Psychology has explored cognition, emotion, and behavior. Yet, despite this accumulation of knowledge, we face an increasing crisis of fragmentation and more. (Wilson, E. O.,1998).

We know more than ever about the parts of the human system, yet we struggle to understand the human being as a whole.

Cognitive Biology is proposed as a unifying framework. It does not reject existing disciplines; instead, it seeks to reconnect them. It asks us to examine how cognition, biological regulation, meaning, and awareness function together as a single living system. (Maturana, H. R., & Varela, F. J., 1980).

Early Life and Biological Conditioning

You already know that every human being enters the world in a state of biological vulnerability. The developing nervous system is highly sensitive to environmental input. Early life experiences play a decisive role in shaping cognitive and biological regulation. (Gunnar, M. R., & Quevedo, K., 2007).

When a child grows in an environment of emotional safety, consistency, and attuned caregiving, the nervous system learns regulation. When the environment is unpredictable, threatening, or emotionally unavailable, the subconsciousness adapts through vigilance and heightened stress reactivity. (Bowlby, J., 1988).

These adaptations are not psychological weaknesses. They are biological survival strategies. However, strategies that ensured survival in early life often persist long after the original environment has changed. Cognitive Biology helps us understand how anxiety, emotional rigidity, and stress-related disorders emerge from outdated biological learning rather than personal failure.

Culture and Belief as Biological Structures

Cultural values, belief systems, moral frameworks, and social narratives further shape cognition. These are not merely intellectual constructs. Through repetition and emotional reinforcement, beliefs become biologically embedded patterns of perception and response.

From a Cognitive Biology perspective, beliefs are stabilized regulatory systems. They influence how stimuli are interpreted, emotions generated, and physiological responses are organised. This understanding explains why deeply held beliefs are resistant to change: altering them requires biological reorganization, not merely logical argument. (Maturana, H. R., & Varela, F. J., 1980).

Personal Philosophy as Embodied Regulation

As individuals mature, experiences accumulate through education, relationships, success, failure, and social interaction. Gradually, these experiences form a personal philosophy of life. This philosophy is not abstract; it is embodied.

Interpretations of life events directly shape stress physiology, emotional regulation, and behavioral tendencies. A person who interprets uncertainty as a threat will exhibit chronic stress activation. A person who interprets uncertainty as a challenge will show adaptive regulation. Behavior, therefore, is the outward expression of internal cognitive-biological organization. (Lazarus, R. S., & Folkman, S., 1984).

Rigidity as a Source of Entropy

Biological systems depend on flexibility. When cognitive frameworks become rigid, biological regulation suffers. Rigid interpretations produce chronic stress. Chronic stress disrupts immune, endocrine, and emotional systems.

Cognitive Biology emphasizes integration rather than opposition. Health is not achieved through control, but through adaptability and balance.

Attention as the Central Regulator

Among all cognitive functions, attention plays a central regulatory role. Attention determines which sensory inputs are processed, which memories are activated, and which emotions are amplified. Biologically, attention directs energy distribution within neural networks. (Posner, M. I., & Petersen, S. E., 1990).

Repeated attention strengthens specific circuits. Attention focused on fear strengthens threat networks. Attention guided by awareness strengthens regulatory pathways. Attention thus serves as a bridge between cognition and physiology.

Regulated Attention and Ethical Qualities

When attention is reactive, behavior becomes impulsive. When attention is regulated, behavior reflects conscious choice. Qualities such as patience, empathy, compassion, and clarity naturally emerge from regulated attention.

These qualities are not merely ethical ideals. They correspond to measurable biological states, including reduced stress hormones, improved immune functioning, and stable emotional regulation. Ethical behaviour, from this perspective, reflects biological coherence. (McEwen, B. S., 1998).

The Mind as a Predictive System

Fundamental attributes of the acquired Mind before the inner transformation are Ego, Greed, Lust, Anger, Attachment, Jealousy, Deceit, Fears, Worries, etc.

The soul, after acquiring the mind, has created the body from a single cell. To graduate to the crown of creation, the human body. As it is connected to absolute intelligence and capable force.

HIERARCHY

SUPER SOUL	UNIVERSAL MIND	BODY OF THE UNIVERSE
SOUL	CAUSAL MIND	CAUSAL BODY (ARDHANAREESHVARA)
SOUL	ASTRAL MIND	ASTRAL BODY
SOUL	PHYSICAL MIND	PHYSICAL BODY

The mind is nothing but information of time and space regarding events and experiences of our life in this material body, and based on that, we pre-calculate the future, fear, and worries, and add all of it as information only. It keeps spinning into past and future aspects. It has laws of nature embedded with it. The most important law of nature is the 3rd law. (As you sow, so shall you reap) Every reaction has an equal and opposite reaction

The physical mind is connected to the next level of the Astral mind. Astral mind has complete information, like a blueprint of various species of life; we have crossed through 8.4 million species. Astral minds have all the information about our past and future of all these lives. Started from amoeba as a single cell, then it dies, the mind again goes back to creative intelligence to the astral region. It dies because all life has a limited lifespan.

For Eg- An RNA and DNA of rats and monkeys are very close to human beings.

Awareness allows individuals to observe predictive patterns rather than being controlled by them. This observation creates space for regulation and biological balance.

Cause, Effect, and Responsibility Without Blame

Biological systems operate according to cause-and-effect principles. Prolonged anger increases cardiovascular strain. Chronic fear suppresses immune function. Sustained calm awareness supports healing.

Cognitive Biology promotes responsibility without blame. Change occurs not by suppressing symptoms, but by restructuring cognition and restoring regulation.

Evolutionary Perspective on Cognition

Cognition is an evolutionary achievement refined over billions of years. From simple organisms to complex nervous systems, life has evolved increasing capacity for regulation, adaptation, and organization. (Darwin, C., 1859).

The human brain reflects principles found throughout nature: feedback, balance, and self-organization. Cognition is not separate from biology; it is biology becoming reflective.

Emotional Regulation and Restraining Intelligence

When cognition operates unconsciously, behavior is reactive. Awareness introduces regulation. Regulation enables choice. This capacity may be described as restraining intelligence: the biological ability to pause, observe, and respond rather than react.

Implications for Mental Health

Many mental health challenges arise from chronic entropy rather than isolated pathology. Anxiety, depression, and emotional instability reflect long-standing imbalances in cognitive-biological systems. (McEwen, B. S., 1998).

Cognitive Biology complements therapeutic approaches by addressing underlying regulatory mechanisms rather than surface symptoms.

Meaning as a Biological Variable

Meaning is often treated as philosophical, but meaning has direct biological consequences. Interpretation determines neural activation patterns and stress responses. Meaning-making is therefore a biological function.

Restructuring meaning becomes a powerful intervention for restoring biological regulation.

Stress and the Biology of Adaptation

Stress is not inherently harmful. It becomes damaging when it is chronic and poorly regulated. Cognitive patterns play a central role in maintaining or resolving stress activation. (Lazarus, R. S., & Folkman, S., 1984).

Awareness interrupts unnecessary biological activation and restores balance.

Beyond Mind–Body Dualism

Cognitive Biology rejects mind–body dualism. Thoughts influence posture, breathing, digestion, and immune responses. Bodily states influence perception, mood, and decision-making.

Cognition is embodied and distributed throughout the organism.

Awareness as an Advanced Biological Capacity

Awareness is not passive. It is an active biological capacity that allows observation without immediate reaction. Repeated awareness strengthens neural circuits associated with regulation, resilience, and emotional stability. (Lutz, A., Slagter, H. A., Dunne, J. D., & Davidson, R. J., 2008).

Implications for Education and Human Development

Learning is enhanced when biological safety is present. Fear inhibits cognition, while safety supports curiosity and engagement. Cognitive Biology encourages educational approaches aligned with biological regulation.

Sustainable Human Transformation

Sustainable transformation occurs when cognition aligns with biological functioning. As awareness increases, regulation improves. As regulation improves, behavior adapts. As adaptation stabilizes, health and well-being emerge. (Damasio, A. R.,1994).

Closing: Toward Integration

Cognitive Biology offers a unified framework for understanding human functioning as a coherent living system. It invites science to move from fragmentation toward integration, from reaction toward regulation, and from survival toward sustainability.

Conclusion

Cognitive biology provides a holistic framework for understanding human life by integrating cognition, biology, meaning, and awareness. It recognises humans as living systems shaped by experience, regulation, and interpretation.

By understanding how cognition shapes biology and how awareness regulates cognition, individuals can move from reactive patterns toward sustainable transformation. This framework has significant implications for psychology, healthcare, education, and human development, offering a unified, biologically grounded approach to human well-being.

References

Maturana, H. R., & Varela, F. J. (1980). *Autopoiesis and Cognition: The Realization of the Living*. Dordrecht: D. Reidel Publishing Company.

Goldstein, E. B. (2019). *Cognitive Psychology: Connecting Mind, Research, and Everyday Experience* (5th ed.). Cengage Learning.

Wilson, E. O. (1998). *Consilience: The Unity of Knowledge.* New York: Alfred A. Knopf. Classic work arguing that excessive specialization fragments knowledge and calling for integration across disciplines.

Gunnar, M. R., & Quevedo, K. (2007). *The neurobiology of stress and development.*

Bowlby, J. (1988). *A secure base: Parent-child attachment and healthy human development. New York, NY: Basic Books.*

Maturana, H. R., & Varela, F. J. (1980). *Autopoiesis and Cognition: The Realization of the Living. Dordrecht: D. Reidel.*

Lazarus, R. S., & Folkman, S. (1984). *Stress, appraisal, and coping. Springer. Establishes that cognitive appraisal (interpretation) of events determines stress responses, not events themselves.*

Posner, M. I., & Petersen, S. E. (1990). *The attention system of the human brain. Annual Review of Neuroscience,* 13, 25–42.

McEwen, B. S. (1998). *Protective and damaging effects of stress mediators. New England Journal of Medicine,* 338(3), 171–179.

Darwin, C. (1859). *On the Origin of Species by Means of Natural Selection.* London: John Murray. Establishes cognition and behavior as outcomes of gradual evolutionary refinement.

Lazarus, R. S., & Folkman, S. (1984). *Stress, appraisal, and coping*. Springer.

McEwen, B. S. (1998). Protective and damaging effects of stress mediators. *New England Journal of Medicine,* 338(3), 171–179.

Lutz, A., Slagter, H. A., Dunne, J. D., & Davidson, R. J. (2008). Attention regulation and monitoring in meditation. *Trends in Cognitive Sciences,* 12(4), 163–169.

Damasio, A. R. (1994). *Descartes' Error: Emotion, Reason, and the Human Brain.* New York: Putnam. Establishes that cognition, emotion, and bodily states are inseparable.

Physical Research, Parapsychology, and Cognitive Biology

Dr. Jerry Solfvin

Associate Professor, University of Massachusetts, Dartmouth, USA

Abstract

The scientific field of parapsychology has a long history of terminological dysfunction, going back to its origins in the Society for Psychical Research in the nineteenth century. There was little agreement upon definitions, taxonomy, and methodology to underpin this new line of scientific research. A new era began with a new name (parapsychology) and the creation of the Duke University Parapsychology Laboratory, where an attempt was made to lay a foundation for the field with a taxonomy, definitions, and standard methodology. This stimulated a great deal of research and brought scientific, media, and public attention to the field in the 1930s. However, the new foundation soon developed cracks and crumbled; the field did not prosper in the long run, as had been hoped, and an agreed-upon answer to the question, "What, exactly, are we studying?" remained unanswered.

Nonetheless, the new era of parapsychology resulted in much published research; peer-reviewed journals that welcomed parapsychological submissions; the founding of the international Parapsychological Association (PA); the acceptance of parapsychology as a member science of the American Association for the Advancement of Science (AAAS); and many academic books, including textbooks, on the topic. Such persistent terminological dysfunction may be a sign or symptom of some deeper issues in the field.

In this paper, I argue that what is most needed right now is NOT another theory but a revisiting of our foundations. This entails re-examining the question, "What, exactly, are we studying?", and scrutinizing our taxonomy and methodological toolbox in light of our progress and developments in science at large. This paper questions whether psi (telepathy, clairvoyance, precognition, PK) is best studied as a psychological phenomenon. It is argued that psi may be more fruitfully framed as *emergent properties of living systems*, including but not limited to humans. The biological sciences - which deal with all living systems, including humans – may provide conceptual and methodological inspiration for studying psi. Biology's two complementary methodological streams – observational (evolutionary) and experimental (functional) – provide a model for balancing experimental vs spontaneous case research. The biological sciences were early to recognize systems thinking as an important (and necessary) tool for studying living processes (such as Bertalanffy's monumental *General Systems Theory*).

Biologists were also early to recognize and embrace complexity and harbored the later development of complex systems science. Importantly, psychology was slow to embrace complexity and complex systems science, and parapsychology has yet to do so. Interestingly, it was biologist/parapsychologist Robert Morris, inaugural Koestler Chair of Parapsychology at Edinburgh University, who advocated for complex systems thinking in parapsychology, pointing out that *psi*, or anomalous experience (AE), as it is more recently called, tends to emerge from complex systems settings (Morris, 1992; 2000). I argue, with Morris, that for parapsychology to move forward, a good beginning would be to embrace the COMPLEXITY of our subject matter. Ideally, we should further agree that *psi* is a product of *open, dynamic, living systems* and may best be researched as a *complex systems science*. This is an important beginning step because it has significant logical sequelae for parapsychology. Morris (2000) wrote, "A system is a set of interactive parts; an open system is a system whose boundaries of influence cannot be precisely delineated.

In parapsychology, our studies become more artificial and sterile as we try to simplify and close off the systems under scrutiny, and it is difficult, if not impossible, to place the researchers guaranteeably outside the system of study." Parapsychology has long been modelled on physical sciences, where linear, quantitative, reductionist, experimental methods have been so successful. Parapsychologists therefore tacitly assume that psi involves a linear, stable state, nondynamic, equilibrium processes. In reference to complexity, Morris points out, "The usual strategies of controlled experimentation, with independent and dependent variables, may often be quite inappropriate. We may need to focus more on strategies for evaluating the output of definable psi conducive systems, foregoing at least at the start the systematic exploration of specific causal linkages."

An important sequela of embracing complexity is to re-enchant parapsychologists with spontaneous case research. In this paper, I argue that the future of our field may hinge upon our ability to make clear, unbiased observations of spontaneous real-life instances of psi. Such observations may provide a logical basis for developing definitions and a taxonomy of case types that go beyond the negative operational definitions of types of psi based upon laboratory procedures and mythology. Pratt & Roll's neologism, *recurrent spontaneous psychokinesis* (RSPK), may be the closest thing we have to a proper, positive, scientific definition of "what, exactly, are we studying?" It fails only because it includes "PK" (psychokinesis), which is a negatively defined theoretical assumption from experimental parapsychology.

In this paper, I propose a variation and extension of the RSPK rubric into an assumption-free, positive definition of a specific type of "complex systems setting" from which psi "tends to emerge" (Morris, 2000). This may also form the basis for a taxonomy of such cases. Specifically, I offer a properly scientific operational definition of a specific complex systems setting from which reliable reports of spontaneous RSPK-type psi phenomena tend to emerge. I call this setting an *anomalous episodic communal kinetic occurrence* (AECKO). AECKO is not a theory or model for psi, but the acronym for a *definition of a specific type of "complex systems setting" from which psi "tends to emerge".* There are certainly other specific "complex systems settings from which psi tends to emerge", which have yet to be identified, defined and catalogued, as Morris (2000) suggested, including experimental and spontaneous settings. **We may need to reframe the notion of taxonomy for parapsychology - a taxonomy NOT based upon types of psi, but types of complex systems settings from which psi tends to emerge.** The implications of embracing complexity and the complex open living systems perspective upon definitions, taxonomy, methodology, and the scientific logic of parapsychology and anomalistic psychology are discussed.

Introduction

Thank you to Dr. Raj Kumar, Prof. Bal Ram Singh, and others involved in organizing this symposium. Let me begin by saying that I am a bit out of my element here because I am neither a biologist nor a Vedic scholar, despite my strong interests in both fields. I am a mathematician and social scientist, most closely associated with the scientific field of parapsychology. The origins of parapsychology date back to 1882 with the founding of the Society for Psychical Research (SPR) in London. I'll need to start with a bit of the history of parapsychology, especially the SPR and the later Duke University Parapsychology Laboratory, primarily to trace the terminological dysfunction that has characterized parapsychology across the years and continues even today. I am referring to the fact that parapsychology historically lacks the terminological foundation usually associated with a field of science – including agreed-upon definitions, taxonomy, and a methodological toolbox in harmony with the terminology. To put it simply, parapsychology does not have an agreed-upon answer to the basic question, "What, exactly, are we studying?" Since I perceive my role as primarily methodological, I take this quite seriously. It is a breach of integrity, a crack in the armor, possibly viewable as a sign or symptom of deeper issues in this field. I also see it as a call to action on behalf of the future of the field. Thus, the terminological dysfunction will be the guiding focus in this brief and cursory dive into the history of parapsychology.

The Society for Psychical Research (SPR)

The scientific field of parapsychology has a long history of terminological dysfunction, going back to its origins in the Society for Psychical Research (SPR) in the nineteenth century. The SPR was founded by a group of distinguished scholars and scientists, many of them quite accomplished. (Gauld, 1968). The mission of the SPR was "to investigate and understand phenomena that appeared to defy conventional scientific explanation. Its primary goal was to examine reports of psychic and paranormal occurrences—such as telepathy, clairvoyance, spiritualism, apparitions, and other related experiences—using rigorous scientific methods." However, there was little agreement on definitions for these targeted phenomena and the methods by which to investigate them. For example, Sir Oliver Lodge believed these phenomena were evidential of an afterlife because they demonstrated information exchange beyond the physical senses. F.W.H. Myers agreed with Lodge concerning prioritizing the afterlife, or "survival" question, but argued that we must first understand the relation of mind and body among the living, and so developed a personality theory of subliminal mind, or subliminal self. French

Nobel Laureate physiologist Charles Richet envisioned "a science" that revealed unknown powers latent in human intelligence. (Mauskopf & McVaugh, 1980, p. 2).

The SPR members set up committees for specific study topics, including telepathy, mesmerism (hypnosis), kinetic/physical phenomena observed around trance mediums during seances, and ghost and haunting reports. One of the SPR projects was the "Census of Hallucinations" (1889-1892), a large-scale study (17,000 respondents) to systematically investigate the frequency and nature of hallucinations in the general population. The study found that such experiences are more widespread than previously thought (about 10% endorsement) and are usually not associated with mental illness. Several so-called "crisis apparitions" were reported, which were suggestive of telepathy, but they yielded no typology and taxonomy (Keep, 2019).

The work of F.W.H. Myers (1903) on post-mortem survival stands out, even today, more than a century later. Myers argued that we must first understand the capacities and limitations of our ***embodied*** human consciousness. Myers was asking, "Is consciousness, like matter, a primary and fundamental characteristic of nature? Or is it secondary …emergent …epiphenomenal …derivative …of nature?" (Cook, 1994; Kelly, Kelly, et al, 2007; Myers, 1903)

Myers postulated a **subliminal self or consciousness** to explain certain living human characteristics or abilities. Myers' research on subliminal consciousness led to the first systematic attempt in parapsychology at a ***taxonomy of human anomalous experience***. These anomalous experiences include telepathy, automatisms, trance states (such as mediumistic trance), apparitions, hypnotic states, multiple personality, genius, and creativity. Myers and the SPR contributed a great deal more to the knowledge base of parapsychology. However, terminological issues were passed on to J.B Rhine and colleagues at the Duke University Parapsychology Laboratory, who, informed by Myers and the SPR, took a very different approach.

Duke University Parapsychology Laboratory

A new era began with a new name (parapsychology) and the creation of the Duke University Parapsychology Laboratory, where an attempt was made to lay a new foundation for the field with a taxonomy, definitions, and standard methodology. In contrast to the Society for Psychical Research (SPR), Rhine envisioned parapsychology as an experimental science to investigate the prospects and limitations of so-called "psychic" experiences among human beings. He focused upon unexplained forms of information exchange, including telepathy (mind to mind), clairvoyance (event to mind), and precognition (future to mind), and called these three *extrasensory perceptions* (ESP). Later, Rhine added psychokinesis (PK, or mind over matter). ESP plus PK is called *psi*. ESP was tested in controlled laboratory settings, such as attempting to "guess" which playing card a remotely located "agent" was simultaneously viewing. Experimental PK testing utilized random processes such as dice rolling, for which the subject would attempt to influence the outcome by thought alone.

Rhine's definitions, however, had a serious problem – they are negative definitions. Rhine defined each of his four forms of psi as "not explainable by known physical processes". You see, "known physical processes" such as sensory cueing, misperception, faulty memory, touching the dice, and random coincidence were rival hypotheses competing with the psi explanation. Rhine's laboratory experiments could eliminate rival hypotheses such as sensory cueing and random chance, was falsified statistically. Repeating such experiments many, many times, with many, many people, Rhine demonstrated apparent information transfer without a known physical explanation.

However, these negative definitions soon proved to be quite problematic. For example, Schouten (1984) wrote, "... a *negative definition*... [for psi] ...does not say what it is, but only what it is not." Hovelmann (1983, p. 131) wrote, "...nothing is explained by making reference to a 'PK force' or to an 'information transfer ' independent of the recognized channels of sense, respectively, as long as we do not know what a 'PK force' is or how the 'information transfer' operates." The powerful American psychologist EG Boring was impressed with the quality of Rhine's careful studies and championed Rhine's efforts. But he advised Rhine not to name this mysterious effect too quickly (e.g., telepathy) because "you have only found *[due to the negative definition]* a very interesting correlation and now you must investigate what that correlation is" (Mauskopf & McVaugh, 1980).

In summary, the Duke Parapsychology Laboratory brought taxonomy, definitions, and experimental methodology to the field. And was able to falsify the random chance hypothesis, which showed there was *something* happening which science could not explain. A peer-reviewed Journal of Parapsychology was created, the Parapsychological Association (PA) was founded, and it was admitted to the American Association for the Advancement of Science (AAAS). The BAD NEWS includes the negative definitions and the lack of repeatability. And we still couldn't answer the question, "What, exactly, are we studying?"

Spontaneous Case Research in Parapsychology

J.B. Rhine was committed to controlled experimental studies, but he did call for more case studies of psi in everyday life in order to provide an additional source of ideas on the planning of experiments. Rhine's wife and colleague, Louisa Rhine, responded by beginning an investigation of the mountains of mail received at the Duke Parapsychology Laboratory, which included reports of personal experiences. Her initial selection criteria was, (LE Rhine, 1977, p.69-70) **"..to accept tentatively for study all reports that were intelligently written and in apparent good faith, if they fulfilled the definition, that is, as reported, they seemed to be instances in which information was received without the senses or effects were produced without the muscles."** Thus, she adopted essentially the same – "negative" - definitions as used for experimental studies. (e.g., "without the senses", and "without the muscles"). She also adopted the same restrictive taxonomy, which was developed for the laboratory – telepathy, clairvoyance, precognition, and psychokinesis.

Louisa Rhine's spontaneous case studies had a much wider impact upon parapsychology than was foreseen. She analyzed thousands of reports, published several books on her results, and many professional journal articles. She developed a typology based upon her findings – intuitive; hallucinatory; unrealistic dreams; and realistic dreams. She demonstrated that qualitative, experience-based data could yield discoveries beyond the reach of controlled laboratory experiments. She found that real-life psi experiences tend to emerge from emotional, meaningful, or urgent circumstances that are nearly impossible to recreate in the laboratory. About half of the reports include a strong sense of "conviction" beyond what's been found in the laboratory, and a sizable proportion of real-life psi reports are death-related. While the spontaneous case analyses of Louisa Rhine and others have added much to the field of parapsychology, it did not improve the fundamental terminological problem that is the focus of the current paper.

Taxonomy

Let's talk about Taxonomy in Science. According to Britannica online, *taxonomy* is derived from the Greek *taxis* (arrangement) and *nomos* (law), and means, in a broad sense, the science of classification. Most usually, taxonomy refers to "the classification of living and extinct organisms-i.e., biological classification", and more specifically "the Linnaean system of binomial nomenclature created by Swedish naturalist Carolus Linnaeus in the 1750s."

According to COPILOT (AI), "A good taxonomy is one of the quiet superpowers of science. It looks like a simple act of naming and sorting, but it actually shapes how scientists **think**, **communicate**, and **discover**." Taxonomy provides a shared language; reveals hidden patterns; guides hypotheses; helps in the accumulation of knowledge; and supports practical applications.

"Taxonomy has been essential to biology because it provides the *organizational backbone* that makes the study of life possible—without it, scientists couldn't reliably identify species, compare organisms, track biodiversity, or build cumulative biological knowledge."

We conclude this brief history of the origins of parapsychology by noting that while Myers produced the first systematic taxonomy of anomalous experience, it was tethered to Myers' subliminal consciousness and subliminal self-theory, and it became one of many that were later proposed. Moreover, proper scientific definitions of these anomalous experiences were often lacking. And there was no agreed-upon answer to the questions, "What, exactly, are we studying, and how?"

Psychology or Life Sciences?

Parapsychology's adoptive methodological parent is the science of psychology, and psychology has its commitment in experimental research, which has been so successful in physics and chemistry, where linear, stable-state, non-adaptive, equilibrium systems are the norm. Much of psychological research meets this requirement, such as perception, cognition, personality, behavior, learning, attitude, emotions, and so forth, in part due to the linear, stable-state, nonadaptive, equilibrium constraints of the research process. This is what biologist Stephen Gould calls the "engineering" approach to science, focused on current utility, efficiency, and mechanisms. Parapsychologists have inherited a preference for this engineering approach and may tacitly assume that psi is the product of linear, stable state, nonadaptive, equilibrium processes.

In this paper, I argue that psi phenomena may be more fruitfully framed as products or *emergent properties of living systems*, including but not limited to humans. The biological sciences - which deal with all living systems, including humans – may provide conceptual and methodological inspiration for the study of psi as a product of living systems. For one thing, the biological sciences are a stark reminder of the importance of a taxonomy that serves as an "*organizational backbone*" and guides and stabilizes a research program. In addition, the biological sciences are built upon two distinct methodological streams – experimental (functional) and observational (evolutionary). This bifurcation maps to parapsychology's experimental and spontaneous case approaches to research, which have long been out of balance, with spontaneous cases being underutilized. Biologists have found a better balance, identifying strengths and weaknesses of both while valuing them equally.

Biologists were early to recognize and embrace complexity and actually harbinged the later development of the field of complex systems science. They were also early to recognize systems thinking as an important (and necessary) tool for studying complex living processes for which experimental methods alone are inadequate.

Living systems are complex, open systems!

Complex living systems cannot be completely understood by experimental methods alone. The study of complex living systems can only be fruitful with methods appropriate to the task.

Interestingly, biologist/parapsychologist Robert Morris, Koestler Chair of Parapsychology at Edinburgh University, advocated for complex systems thinking in parapsychology. Morris (1992; 2000) pointed out that *psi* tends to emerge from complex systems settings. Morris (2000) further explained, "A system is a set of interactive parts; an open system is a system whose boundaries of influence cannot be precisely delineated. In parapsychology, our studies become more artificial and sterile as we try to simplify and close off the systems under scrutiny, and it is difficult, if not impossible, to place the researchers guaranteeably outside the system of study."

Morris points out something that is now widely known about complexity: "The usual strategies of controlled experimentation, with independent and dependent variables, may often be quite inappropriate." Morris continues, "We may need to focus more on strategies for evaluating the output of definable psi-conducive systems, foregoing, at least at the start, the systematic exploration of specific causal linkages."

Morris is here suggesting a reframe, two, actually. If the "usual strategies of controlled experimentation, with independent and dependent variables" are inappropriate, spontaneous case research becomes a more attractive alternative. The second, very important, reframe is: switching from **psi-conducive states**, which emphasizes individual person, to psi-conducive settings, which de-emphasize the individual, in favor of the past, present, and future factors that modulate the current setting.

With Morris, I advocate for parapsychologists to embrace the complexity of psi phenomena. Complexity science makes it all the more urgent to go beyond the negative definitions of psi phenomena that we still cling to. Pratt & Roll's simple, elegant neologism (Pratt & Roll, 1957), *recurrent spontaneous psychokinesis* (RSPK), comes very close to fulfilling Morris's implied taxonomic criteria of "complex systems setting" from which psi "tends to emerge" (Morris, 2000). It fails mainly due to the inclusion of negatively defined "PK" (psychokinesis).

Building upon this, Bryan Williams and I have updated it into a proper scientific operational definition of a specific complex systems setting from which reliable reports of spontaneous psi tend to emerge. We call this setting an *anomalous episodic communal kinetic occurrence* (AECKO). AECKO is not a theory or model for psi, but the acronym for a *definition of a specific type of "complex systems setting" from which psi "tends to emerge".* The acronym AECKO is defined this way: it is anchored at the ends by "A" and "O", for anomalous occurrence, a BURST or repetition of several (2+) anomalies. The "E" stands for EPISODIC, meaning a narrative, story, or episode of a story, which is bounded by time and space, meaning that it has a beginning and an ending, with events taking place in a particular boundary location, usually a family residence. The "C" is for communal, the group of people, witnesses, reporters of it, and victims that form a "family" of sorts brought together by this EPISODE. And, finally, "K" is for kinetics, the energetic effects such as sounds and object movements that are a key identifying feature for AECKO cases.

Conclusion

Starting from the identification of terminological dysfunction in the field of parapsychology, going back to its beginnings in the Society for Psychical Research (SPR), continuing through and beyond the Duke University Parapsychology Laboratory, we arrive at the fact that this field has no agreed-upon answer to the basic question, What, exactly, are we studying? I argue that psi (telepathy, clairvoyance, precognition, psychokinesis) should not be assumed to be anomalies of MIND, as they are now, but rather, products of complex, open, living systems. The life sciences may offer better models, means, and methods for studying psi. To proceed, we must embrace the complexity of psi. I advocate, with Morris (1992, 2000), for reframing the taxonomy for psi, not based on types or causes, but upon types of complex systems settings from which psi tends to emerge. I offer one complex system setting (AECKO) from which psi tends to emerge as an example for adding other settings to this reframed taxonomy of complex systems settings from which psi tends to emerge.

Acknowledgements

Many thanks to Bryan Williams for his supportive comments and insights during the development of this study. Parts of this study were supported by the BIAL Foundation under Grant No. 387/20.

References

Cook, Emily Williams. (1994). The subliminal consciousness: F.W.H. Myers's approach to the problem of survival. *Journal of Parapsychology, 58*, 39-58.

Gauld, A. (1968). *The founders of psychical research.* Routledge & Kegan Paul

Hovelmann, G. (1983). Seven recommendations for the future of parapsychology. *Zetetic Scholar, 11*, 128-138.

Keep, Christopher. (2019). Evidence in matters extraordinary: Numbers, narratives, and the Census of Hallucinations. *Victorian Studies, 61*(4), 582-607.

Kelly, E.F., Kelly, E.W., Crabtree, A., Gauld, A., Grosso, M., & Greyson, B. (2007). *Irreducible Mind: Toward a psychology for the 21st century.* Lanham, MD: Rowman & Littlefield.

Lucadou, W.V. (1984). What is wrong with the definition of psi? *European Journal of Parapsychology, 5*, 261-283.

Mauskopf, S.H., & McVaugh, M.R. (1980). *The elusive science: Origins of experimental psychical research.* Baltimore: Johns Hopkins University Press.

Morris, R.L. (1992). Parapsychology in the 1990's: Addressing the challenge. *European Journal of Parapsychology, 8*, 1-26.

Morris, R. L. (2000). Parapsychology in the 21st century. *Journal of Parapsychology*, *64*(2), 123-138.

Myers, F.W.H. (1903). *Human Personality and its Survival of Bodily Death. (2 Vols.) London: Longman, Greens.*

Pratt, J.G., & Roll, W.G. (1958). The Seaford disturbances. *Journal of Parapsychology, 22*, 79-124.

Rhine, L.E. (1976). Psi, what is it? The story of ESP and PK: An introduction to parapsychology. NY: Harper and Rowe.

Schouten, Sybo. (1983). A different approach for analyzing spontaneous cases. *Journal of Parapsychology, 47*, 323-340.

Willin, M. (2021). Alan Gauld. Psi Encyclopedia. The Society for Psychical Research.

Delaying Cognitive Decline in Alzheimer's Disease: A Comprehensive Framework Integrating Primary, Secondary, and Tertiary Prevention with Disease-Modifying Therapies, Neurorehabilitation, Neuromodulation, Brain–Computer Interfaces, and Artificial Intelligence

Dr. Nirmal Surya

Chairman, Surya Neuro Center, Mumbai, India; President, Indian Federation of Neuro Rehabilitation; President, Asian Oceanian Society of Neuro Rehabilitation; Secretary General, World Federation of Neurorehabilitation

Abstract

Alzheimer's disease (AD) is a progressive neurodegenerative disorder in which pathological changes—including amyloid-β plaque accumulation, tau neurofibrillary tangle formation, and neuroinflammation—precede clinical symptom onset by decades (Jack et al., *Lancet Neurology*, 2010; Sperling et al., *Alzheimer's & Dementia*, 2011). This review synthesizes the contemporary evidence for a comprehensive, stage-specific framework to delay cognitive decline across the AD continuum. **Primary prevention** targets cognitively normal individuals with preclinical biomarker evidence through modifiable lifestyle risk factor optimization, cardiovascular health management, and emerging preclinical trials such as the AHEAD Study. **Secondary prevention** encompasses anti-amyloid disease-modifying therapies (DMTs)—lecanemab and donanemab—which have demonstrated 27–35% slowing of cognitive and functional decline in landmark phase III trials (van Dyck et al., *NEJM*, 2023; Sims et al., *JAMA*, 2023). **Tertiary management** integrates symptomatic pharmacotherapy with cognitive rehabilitation, physical exercise, psychosocial interventions, and lifestyle modification. Beyond pharmacotherapy, this review presents a paradigm-shifting argument for rehabilitation as a biologically active, disease-modifying intervention through its effects on neuroplasticity, cognitive reserve, and network reorganization. The integration of repetitive transcranial magnetic stimulation (rTMS), brain–computer interfaces (BCIs), and artificial intelligence (AI) represents the frontier of precision neurorehabilitation, with combined approaches potentially achieving 10–12+ months of cognitive delay—substantially exceeding any single modality alone. The evidence reviewed herein supports a transition from passive dementia management to active, sustained, technology-enhanced engagement of neural networks across the disease continuum.

Keywords: Alzheimer's disease; cognitive decline; disease-modifying therapy; lecanemab; donanemab; neurorehabilitation; cognitive reserve; rTMS; brain–computer interfaces; artificial intelligence; neuroplasticity; ARIA; amyloid; tau; precision medicine

Introduction: When Does Cognitive Decline Really Begin?

Alzheimer's disease represents the most significant neurodegenerative cause of cognitive impairment worldwide, contributing substantially to years lived with disability and imposing an immense burden on individuals, families, healthcare systems, and economies (GBD 2019 Dementia Forecasting Collaborators, *Lancet Public Health*, 2022). The critical insight reshaping contemporary understanding of AD is that neurodegenerative changes begin silently—often decades before the first clinical symptoms become apparent (Jack et al., 2010; Sperling et al., 2011). By the time memory failure is noticeable, the underlying disease biology—including amyloid-β plaque deposition, phosphorylated tau accumulation, and progressive synaptic loss—is already firmly established in brain tissue.

This temporal dissociation between pathological initiation and clinical manifestation represents both a diagnostic challenge and a therapeutic opportunity. Critical windows for early intervention are frequently overlooked due to the absence of visible symptoms, creating missed opportunities for disease trajectory modification. Even modest delays in symptom onset translate into meaningful functional preservation, allowing individuals to maintain their dignity, relationships, and daily routines longer. The benefits extend beyond the individual—delaying decline has profound family, economic, and societal implications (Livingston et al., *Lancet*, 2020).

Understanding dementia as a gradual continuum rather than a precipitous event fundamentally changes our approach to intervention and care. Neurodegeneration unfolds progressively across the entire lifespan; clinical decline reflects the cumulative accumulation of biological injury over many years; and numerous intervention opportunities exist throughout the disease course—from risk factor modification to active rehabilitation (Dubois et al., *Lancet Neurology*, 2016). Single, late-stage interventions alone are insufficient to meaningfully alter the disease trajectory. This review presents a comprehensive, evidence-based framework for delaying cognitive decline that integrates primary, secondary, and tertiary prevention strategies with cutting-edge neurorehabilitation, neuromodulation, and computational technologies.

The Biological Landscape: Pathophysiology and the ATN Framework

Amyloid–Tau–Neurodegeneration Cascade

The ATN classification framework—amyloid (A), tau (T), and neurodegeneration (N)—provides the biological architecture for understanding AD as a continuum rather than a clinical syndrome alone (Jack et al., *Alzheimer's & Dementia*, 2018). Amyloid-β accumulation occurs in the preclinical stage, an estimated 15–20 years before symptom onset, establishing the pathological substrate upon which subsequent tau pathology and neurodegeneration unfold. Phosphorylated tau biomarkers—particularly p-tau 181 and p-tau 217—correlate closely with the cognitive decline trajectory, while neurofilament light chain (NfL) elevation serves as a biomarker of ongoing axonal damage and progressive synaptic loss (Palmqvist et al., *JAMA*, 2020).

The concept of cognitive reserve—the brain's adaptive capacity to maintain function despite accumulating neuropathology—explains why identical pathological burdens produce vastly different clinical presentations across individuals. Education, occupational complexity, and lifelong learning build neural redundancy that delays clinical symptom expression, operating independently of disease prevention itself (Stern, *Lancet Neurology*, 2012). This buffering effect has profound implications: it modulates individual response to therapeutics, informs prognosis and care planning, and critically justifies cognitive engagement interventions at every disease stage.

Mixed Pathology and Vascular Contributions

The majority of dementia cases do not present as pure Alzheimer's disease. Instead, they represent a convergence of AD pathology and cerebral vascular disease, complicating both diagnosis and therapeutic approaches. Cerebral small vessel disease independently accelerates cognitive decline, acting as a critical modifier of disease trajectory through white matter hyperintensities and lacunar infarcts that progressively erode cognitive reserve (Dichgans & Leys, *Lancet Neurology*, 2017). The SPRINT-MIND trial demonstrated that aggressive systolic blood pressure control to <120 mmHg significantly reduced the risk of mild cognitive impairment, establishing vascular risk management as a disease-modifying strategy (Williamson et al., *JAMA*, 2019).

Neuroinflammation: The Engine of Ongoing Decline

While amyloid deposition may initiate pathology, sustained microglial activation drives the ongoing neurodegeneration that correlates most strongly with clinical progression. Translocator protein positron emission tomography (TSPO-PET) studies have demonstrated that microglial activation correlates with disease progression more reliably than amyloid burden alone (Hamelin et al., *Brain*, 2016). Sustained neuroinflammation directly drives synaptic loss and network disconnection, connects neurodegeneration with systemic inflammatory conditions and metabolic dysfunction, and represents a critical focus for next-generation disease-modifying interventions. The inflammatory cascade, rather than protein deposition *per se*, may represent the most actionable therapeutic window for slowing functional decline.

Table 1. ATN Framework Biomarkers and Their Clinical Significance in Alzheimer's Disease

ATN Category	Key Biomarkers	Temporal Onset	Clinical Significance
Amyloid (A)	*Aβ*42/40 ratio, Amyloid PET, p-tau 217	15–20 years preclinical	Initiates pathological cascade; mandatory for DMT eligibility
Tau (T)	p-tau 181, p-tau 217, Tau PET	5–15 years preclinical	Correlates with cognitive decline trajectory; stratifies DMT response
Neurodegeneration (N)	NfL, hippocampal volume (MRI), FDG-PET	Overlaps with early symptoms	Indicates irreversible damage; prognostic of rapid decline
Inflammation	TSPO-PET, sTREM2, GFAP	Throughout the disease course	Drives ongoing neurodegeneration; emerging therapeutic target

Abbreviations: *Aβ, amyloid-beta; NfL, neurofilament light chain; FDG, fluorodeoxyglucose; TSPO, translocator protein; sTREM2, soluble triggering receptor expressed on myeloid cells 2; GFAP, glial fibrillary acidic protein. Adapted from Jack et al. (2018) and Palmqvist et al. (2020).*

Primary Prevention: Targeting the Preclinical Stage

Primary prevention targets cognitively normal individuals who harbour preclinical biomarker evidence of AD pathology. Non-invasive blood biomarker screening—including plasma p-tau 217, p-tau 181, phospho-tau 217/181, and NfL—now enables risk stratification in community settings without the cost and invasiveness of amyloid PET or cerebrospinal fluid analysis (Hansson et al., *Lancet Neurology*, 2022). APOE ε4 genotyping further refines risk: homozygosity for APOE ε4 confers an approximately 3-fold increased lifetime risk of developing AD (Belloy et al., *Nature Medicine*, 2024).

Modifiable Lifestyle Risk Factors

The 2020 Lancet Commission on Dementia identified 12 modifiable risk factors that collectively account for approximately 40% of worldwide dementias (Livingston et al., 2020). Physical exercise (≥150 min/week aerobic plus resistance training) increases brain-derived neurotrophic factor (BDNF), preserves hippocampal volume, and improves cerebral perfusion. Cognitive stimulation through learning, puzzles, reading, and social engagement promotes synaptic plasticity and cognitive reserve accumulation. Cardiovascular health optimization—blood pressure control, lipid management, and diabetes prevention—reduces cerebral amyloid burden and provides direct vascular neuroprotection.

Sleep as a Biological Disease Modifier

During slow-wave sleep, the glymphatic system actively clears amyloid-β and tau proteins from the brain parenchyma (Xie et al., *Science*, 2013). Sleep disorders disrupt this critical clearance pathway, accelerating pathological protein accumulation and cognitive decline. Treatment of obstructive sleep apnoea demonstrably improves cognitive trajectory and slows disease progression (Leng et al., *Nature Reviews Neurology*, 2023). Sleep optimization is not merely supportive care—it is biologically disease-modifying.

Preclinical Intervention Trials

The AHEAD 3-45 Study (NIH Phase III) is evaluating lecanemab in preclinical AD—cognitively normal individuals with confirmed amyloid positivity. Results are anticipated in 2025–2026 and may extend the indication for anti-amyloid DMTs to asymptomatic amyloid-positive individuals, opening a transformative new era of truly presymptomatic intervention. The estimated timeline from biomarker positivity to cognitive symptoms is approximately 10–15 years, representing an extended therapeutic window amenable to primary prevention strategies.

Table 2. Modifiable Risk Factors for Cognitive Decline and Their Biological Mechanisms

Risk Factor Domain	Recommended Intervention	Biological Mechanism
Physical Exercise	≥150 min/week aerobic + resistance training	BDNF upregulation, hippocampal volume preservation, cerebral perfusion improvement
Cognitive Stimulation	Learning, puzzles, reading, social engagement	Synaptic plasticity enhancement, cognitive reserve accumulation
Cardiovascular Health	BP control, lipid management, diabetes prevention	Reduced cerebral amyloid burden, vascular neuroprotection
Sleep Optimization	7–9 hours nightly; OSA screening and treatment	Glymphatic clearance of Aβ and tau during slow-wave sleep
Diet Quality	Mediterranean/DASH diet; omega-3, antioxidants	Neuroinflammation reduction, amyloid burden modulation
Social Engagement	Group activities, community involvement, peer support	Cognitive reserve building, depression prevention

Abbreviations: *BDNF, brain-derived neurotrophic factor; BP, blood pressure; OSA, obstructive sleep apnoea; Aβ, amyloid-beta. Adapted from Livingston et al. (Lancet, 2020).*

Secondary Prevention: Disease-Modifying Therapies

Anti-Amyloid Monoclonal Antibodies

Disease-modifying therapies for AD are specifically indicated for patients in the early stages of the disease continuum: mild cognitive impairment (MCI) due to AD and mild AD dementia with biomarker-confirmed amyloid pathology. Biological confirmation through amyloid PET imaging or validated plasma biomarkers (Aβ42/40 ratio, p-tau 217) is mandatory before DMT initiation, in accordance with the NIA–AA biological research framework (Jack et al., 2018).

Lecanemab (Leqembi®)

Lecanemab is a humanized IgG1 monoclonal antibody that selectively targets aggregated soluble amyloid-β protofibrils and insoluble amyloid plaques. The Clarity AD trial (N=1,795; 18 months) demonstrated statistically significant slowing of decline across all co-primary and key secondary endpoints (van Dyck et al., *NEJM*, 2023). The primary endpoint, CDR-SB at 18 months, showed a 27% slowing of cognitive and functional decline versus placebo (0.45-point absolute benefit; $P<0.0001$). ADAS-Cog14 demonstrated a 26% slowing of cognitive decline ($P<0.001$), ADCS-MCI-ADL showed a 37% slowing of functional decline ($P<0.0001$), and 68% of treated patients achieved amyloid plaque clearance below the 30-Centiloid threshold ($P<0.0001$). Lecanemab received FDA traditional approval in July 2023 and EMA authorization in 2024. Dosing is 10 mg/kg intravenously biweekly, with transitioning to monthly dosing after 18 months supported by pharmacokinetic modelling.

Donanemab (Kisunla®)

Donanemab targets deposited N-terminal pyroglutamate-modified amyloid-β. The TRAILBLAZER-ALZ 2 trial (N=1,736; 76 weeks) achieved its primary endpoint with a 35% slowing of cognitive and functional decline on the iADRS versus placebo, with CDR-SB demonstrating a 25% reduction in cognitive decline progression (Sims et al., *JAMA*, 2023). A notable subgroup advantage was observed in participants with low/medium tau burden, supporting tau-stratified patient selection. Donanemab offers a dosing convenience advantage with monthly intravenous administration and a unique treatment-to-target paradigm that allows discontinuation after amyloid clearance is confirmed by PET. FDA approval was granted in July 2024.

Table 3. Comparative Characteristics of Lecanemab and Donanemab

Characteristic	Lecanemab (Leqembi®)	Donanemab (Kisunla®)
Molecular Target	Soluble Aβ protofibrils + plaques	Deposited N3pG-modified Aβ
Pivotal Trial	Clarity AD (N=1,795; 18 months)	TRAILBLAZER-ALZ 2 (N=1,736; 76 wk.)
Primary Endpoint	CDR-SB: 27% slowing (P<0.0001)	iADRS: 35% slowing (P<0.001)
CDR-SB Benefit	0.45-point absolute benefit at 18 mo.	25% slowing of CDR-SB decline
Dosing Regimen	10 mg/kg IV biweekly (continuous)	Weight-tiered IV monthly (stop at clearance)
ARIA-E Incidence	~12.6%	~17.3% (24% in some analyses)
ARIA-H Incidence	~17.3%	~12.7%
Regulatory Status	FDA 2023; EMA 2024	FDA July 2024
Clinical Interpretation	~4–6-month delay in progression	~4–6-month delay; greater in low-tau

Abbreviations: *Aβ, amyloid-beta; CDR-SB, Clinical Dementia Rating–Sum of Boxes; iADRS, Integrated Alzheimer's Disease Rating Scale; ARIA-E, amyloid-related imaging abnormalities–edema; ARIA-H, amyloid-related imaging abnormalities–hemorrhage.* ***Sources:*** *van Dyck et al. (NEJM, 2023); Sims et al. (JAMA, 2023); Nature Reviews Neurology comparative analyses (2024).*

ARIA: Recognition, Monitoring, and Management

Amyloid-related imaging abnormalities represent the principal safety concern with anti-amyloid monoclonal antibodies. ARIA encompasses both vasogenic edema/effusion (ARIA-E) and cerebral microhaemorrhages/superficial siderosis (ARIA-H). Most ARIA events are radiographic findings without symptoms; when symptomatic, presentations include headache, confusion, visual disturbances, focal neurological deficits, or seizures. APOE ε4 homozygosity is associated with substantially higher ARIA rates—

approximately 15% ARIA-E in ε4 homozygotes compared to 8% in heterozygotes in the lecanemab trial (van Dyck et al., 2023). Monitoring mandates serial brain MRI with FLAIR and gradient-echo sequences before infusions 5, 7, 9, and 14 for lecanemab, with additional imaging for any new neurological symptoms. Management follows the ARIA Consensus Guidelines published in *JAMA Neurology*, employing dose hold, corticosteroid therapy for symptomatic cases, and permanent discontinuation for severe or recurrent events.

Table 4. ARIA Recognition and Management Protocol

Step	Action	Clinical Consideration
1. Clinical Presentation	Most cases asymptomatic; symptomatic: headache, confusion, visual changes, seizures	Severity ranges from mild to life-threatening; low threshold for imaging
2. Detection & Monitoring	Serial MRI (FLAIR + GRE/SWI) per protocol schedule; urgent MRI if new symptoms	Peak risk during the first 6 months of therapy; APOE ε4 status informs frequency
3. Acute Management	Immediate dose hold; urgent brain MRI; corticosteroids for symptomatic ARIA	Severe cases require hospitalisation and neurological consultation
4. Treatment Resumption	Mild asymptomatic ARIA-E: cautious resumption after radiographic resolution	Severe/recurrent/symptomatic ARIA: permanent discontinuation is typically warranted

Abbreviations: *MRI, magnetic resonance imaging; FLAIR, fluid-attenuated inversion recovery; GRE, gradient-recalled echo; SWI, susceptibility-weighted imaging. Adapted from JAMA Neurology ARIA Consensus Guidelines.*

Clinical Implementation: From Syndrome to Therapy

The pathway to DMT initiation requires systematic clinical phenotyping (prominent episodic memory impairment with gradual onset), biological confirmation via amyloid PET or plasma biomarkers, safety screening with baseline brain MRI (GRE/SWI sequences), and structured shared decision-making with patients and care partners regarding realistic expectations, monitoring requirements, ARIA risks, and the modest magnitude of clinical benefit. Eligibility requires age 50–85 years, MMSE 20–26, CDR-SB 1.5–7.0, and confirmed amyloid positivity. Exclusion criteria include more than four cerebral microbleeds on baseline MRI, prior intracerebral haemorrhage, extensive superficial siderosis, and current anticoagulation therapy.

Tertiary Management: Symptomatic Therapy, Rehabilitation, and Lifestyle

Symptomatic Pharmacotherapy

In moderate-to-advanced AD (CDR-SB ≥5), cholinesterase inhibitors (donepezil 5–10 mg, rivastigmine, galantamine) provide modest 6–12-month cognitive benefit through acetylcholine augmentation. The NMDA receptor antagonist memantine (20 mg) confers moderate benefit in moderate-to-severe AD through reduction of glutamatergic excitotoxicity. Combination therapy with donepezil plus memantine demonstrates superiority over monotherapy, with effect sizes of 0.5–0.8 standard deviations on cognitive scales (Tariot et al., *JAMA*, 2004). The synergy concept—combining DMTs (disease modification) with cholinesterase inhibitors and memantine (symptomatic benefit)—represents an additive cognitive preservation strategy that exceeds either approach alone.

Cognitive Rehabilitation as Biological Therapy

For decades, rehabilitation in cognitive disorders was positioned as purely supportive—something offered after the disease declared itself. This framing is fundamentally outdated. Contemporary neuroscience demonstrates that rehabilitation can influence the very biological processes that drive cognitive decline by actively shaping neural networks, slowing functional deterioration, and modifying the clinical trajectory of disease (Clare et al., *Cochrane Database of Systematic Reviews*, 2023).

Cognitive stimulation therapy (CST) consists of structured group sessions targeting memory, attention, and language domains, achieving effect sizes of approximately 0.5 with a 6-month cognitive delay benefit. Cognitive rehabilitation (CRT) employs personalised strategy training—mnemonic techniques, spaced retrieval, external aids—with a focus on real-world functional outcomes rather than task-specific performance. The neuroplasticity principle underpinning these approaches is that intensive, progressively challenging, multimodal training enhances synaptic compensation even in the presence of ongoing neurodegeneration.

Crucially, neuroplasticity does not disappear in neurodegenerative disease. In MCI and early AD, synaptic modification, learning-dependent reorganisation, and experience-driven structural changes remain surprisingly robust (Vemuri et al., *Neurology*, 2012). The presence of pathology does not eliminate the brain's capacity to adapt—it simply requires more strategic and sustained engagement to activate plasticity mechanisms effectively. Rehabilitation induces neuroplasticity through long-term potentiation of synaptic connections, enhanced network synchronisation across distributed brain regions, and upregulation of BDNF and other neurotrophic factors that support neuronal survival and structural remodelling.

Physical Exercise as Cognitive Therapy

Physical exercise is itself a form of biological cognitive therapy with direct effects on brain structure and function. Multi-modal exercise programmes combining aerobic, resistance, and balance training achieve the largest standardised mean differences (SMD 5.58) in cognitive outcomes, compared to aerobic exercise alone (SMD 4.22) or cognitive training alone (SMD 3.85) (Northey et al., *British Journal of Sports Medicine*, 2018). The mechanisms are well-characterised: aerobic exercise stimulates hippocampal neurogenesis through BDNF upregulation, improves cerebral blood flow and angiogenesis, and reduces vascular risk factors. Resistance training independently strengthens prefrontal networks governing executive control. Combined aerobic and resistance programmes optimise cognitive outcomes—the optimal dose being 150 min/week aerobic plus twice-weekly resistance training, supplemented by cognitive training three or more sessions per week, supervised where possible.

Table 5. Comparative Efficacy of Cognitive Delay Interventions

Intervention	Estimated Delay	Primary Mechanism	Evidence Level
DMT Monotherapy	4–6 months	Amyloid burden reduction	Phase III RCTs (Level I)
Cognitive Rehabilitation Alone	~6 months	Synaptic plasticity enhancement	RCTs/Meta-analyses (Level I–II)
Physical Exercise (Multimodal)	4–6 months	BDNF, vascular plasticity, and neuroinflammation reduction	RCTs/Meta-analyses (Level I)
Combined DMT + Rehab + Exercise + Lifestyle	10–12+ months	Synergistic neuroprotection across multiple pathways	Extrapolated (Level II–III)
DMT + Anti-Tau + Rehab (Future)	12–18+ months	Multi-mechanistic disease modification	Projected (Phase II–III pending)

Abbreviations: *DMT, disease-modifying therapy; BDNF, brain-derived neurotrophic factor; RCT, randomised controlled trial. Effect sizes derived from Clarity AD, TRAILBLAZER-ALZ 2, FINGER trial, and meta-analytic data.*

Repetitive Transcranial Magnetic Stimulation: From Symptomatic Relief to Disease Modification

Neuromodulation with rTMS addresses the fundamental limitation of behavioural rehabilitation—its dependence on endogenous plasticity that may be compromised in neurodegenerative disease. rTMS externally induces plasticity through well-characterised neurophysiological mechanisms: frequency-dependent modulation of cortical excitability (high-frequency ≥5 Hz increases excitability; low-frequency ≤1 Hz decreases it), induction of long-term potentiation/depression-like synaptic changes, and reorganisation of functional connectivity within and between neural networks (Lefaucheur et al., *Clinical Neurophysiology*, 2020).

From Anatomical to Network-Targeted rTMS

First-generation rTMS trials used anatomically defined targets—typically the dorsolateral prefrontal cortex—yielding inconsistent results and modest effect sizes. This approach failed because it did not account for network-level dysfunction, individual variability in connectivity patterns, or the distributed nature of cognitive processing. Network-targeted rTMS represents a fundamental methodological advance: it uses functional MRI connectivity data to identify personalised cortical targets that optimise engagement of specific cognitive networks in each patient (Koch et al., *Brain*, 2022).

Hippocampal Network and Precuneus rTMS

Although the hippocampus is inaccessible to direct TMS stimulation due to its depth, its network extends to TMS-accessible cortical regions, particularly in parietal cortex. High-frequency rTMS delivered to individualised parietal targets improves episodic memory performance and increases hippocampal–parietal functional connectivity in patients with MCI and early AD (Wang et al., *Brain*, 2014). The precuneus, a critical hub within the default mode network showing early vulnerability in AD, represents another promising target. Network-targeted precuneus rTMS over 20 sessions combined with concurrent cognitive rehabilitation has shown sustained cognitive benefits at 12-month follow-up—suggesting genuine disease modification rather than transient symptomatic improvement (Koch et al., 2022).

Synergy: rTMS Plus Rehabilitation

The combination of rTMS and cognitive rehabilitation produces effects significantly exceeding either intervention alone. This synergy reflects their complementary mechanisms: rTMS creates a state of enhanced plasticity, while rehabilitation directs that plasticity toward functionally relevant learning. Timing is critical; delivering rTMS immediately before or during rehabilitation sessions maximises the therapeutic window when networks are primed for learning. This temporal pairing transforms passive stimulation into active learning enhancement, effectively converting rehabilitation into biologically active therapy with potential disease-modifying properties.

Brain–Computer Interfaces in Cognitive Rehabilitation

A brain–computer interface enables direct communication between neural activity and an external device by decoding brain signals in real time and translating them into actionable outputs. In cognitive disorders, behavioural responses become increasingly unreliable indicators of underlying brain state—patients may struggle to execute responses despite intact neural processing. BCIs bypass this behavioural bottleneck by accessing neural signals directly, enabling continuous brain state monitoring, adaptive intervention delivery, and early disengagement detection (Mane et al., *Journal of Neural Engineering*, 2020).

BCIs in cognitive rehabilitation function not as assistive devices replacing lost function, but as therapeutic neuromodulation systems that actively reshape neural networks through operant conditioning at the neural level. When specific neural patterns trigger immediate feedback, the brain learns to produce those patterns more reliably, enhancing attention-related theta activity, strengthening network connectivity, and improving sustained cognitive engagement. Closed-loop BCI systems continuously detect cognitive performance, fatigue, and attention levels, then dynamically adjust task demands to maintain optimal challenge—preventing both undertraining and overtraining.

Despite their promise, current BCIs face limitations including signal quality challenges, inter-individual variability necessitating personalised calibration, technical complexity limiting accessibility, and scalability constraints. These limitations highlight the critical need for integration with artificial intelligence systems capable of handling the complexity that prevents current BCIs from reaching their therapeutic potential.

Artificial Intelligence in Dementia Rehabilitation

Cognitive decline is profoundly heterogeneous—no two patients follow identical trajectories or respond identically to interventions. This heterogeneity creates complexity exceeding human cognitive capacity to manage optimally, making AI essential infrastructure rather than an optional enhancement. AI transforms dementia care across four critical domains: cognitive assessment through continuous passive monitoring of speech patterns, gait dynamics, digital behaviour, and daily activity (Buckley et al., *Alzheimer's & Dementia*, 2022); personalisation of rehabilitation through real-time performance profiling and dynamic task selection; integration with rTMS to identify personalised stimulation targets and predict treatment response; and integration with BCIs to extract meaningful patterns from noisy EEG signals and enable robust clinical deployment.

Perhaps AI's most transformative impact is enabling scalability. Dementia affects millions globally, but intensive rehabilitation requires specialised expertise that is scarce and geographically concentrated. AI-powered systems enable remote monitoring, continuous adaptation, and long-term engagement without requiring constant clinician supervision—translating rehabilitation from boutique intervention to standard care. The transition from static rehabilitation protocols to adaptive algorithms that respond to changes in brain state, behaviour, and disease stage represents a fundamental evolution in dementia management. Ethical considerations—including clinical governance, algorithmic transparency, data privacy, and equitable access—must be addressed proactively as AI is integrated into clinical practice.

The Integrated Rehabilitation Model: A Unified Framework

The central thesis of this review is that delaying cognitive decline requires moving beyond single interventions toward integrated therapeutic frameworks. The future of dementia care lies in combining cognitive rehabilitation, rTMS neuromodulation, brain–computer interfaces, and artificial intelligence into unified, synergistic systems where each component addresses the limitations of the others.

Cognitive rehabilitation provides structured learning and strategy development. rTMS amplifies neuroplasticity and modulates network function. BCIs enable real-time brain state monitoring and adaptive feedback. AI delivers personalisation, prediction, and scalable deployment. Together, combined DMT + symptomatic therapy + cognitive rehabilitation + exercise + lifestyle optimisation achieves an estimated 10–12+ months of cognitive delay—substantially exceeding the 4–6-month benefit of any single modality. This stage-specific approach acknowledges that therapeutic goals shift as disease progresses: in MCI, intensive plasticity-based interventions aim to delay conversion; in early dementia, multimodal rehabilitation with neuromodulation modifies trajectory; in moderate-to-advanced disease, functional maintenance and caregiver-supported engagement preserve quality of life.

Table 6. Stage-Specific Integration of Therapeutic Modalities Across the AD Continuum

Disease Stage	Primary Interventions	Technology Integration	Therapeutic Goal
Preclinical Risk	Lifestyle modification, cardiovascular health, cognitive engagement	AI-driven risk screening, digital biomarkers	Reserve building, risk reduction
MCI (Prodromal AD)	DMTs, intensive cognitive rehabilitation, exercise	AI-personalised rehab, rTMS plasticity induction, BCI neurofeedback	Delay conversion to dementia
Early AD Dementia	DMTs, combined rehab + neuromodulation, symptomatic Rx	Network-targeted rTMS, closed-loop BCI, AI adaptation	Trajectory modification, independence preservation
Moderate–Advanced	Symptomatic pharmacotherapy, functional rehabilitation, caregiver support	BCI engagement maintenance, AI-guided caregiver training	Functional preservation, slow dependency

Abbreviations: *MCI, mild cognitive impairment; AD, Alzheimer's disease; DMT, disease-modifying therapy; rTMS, repetitive transcranial magnetic stimulation; BCI, brain–computer interface; AI, artificial intelligence; Rx, pharmacotherapy.*

Future Directions

The therapeutic landscape for AD is evolving rapidly toward multi-mechanistic approaches. Anti-tau monoclonal antibodies targeting p-tau 217 are in Phase III trials, with anticipated 20–25% additional cognitive decline slowing. Dual anti-amyloid plus anti-tau strategies hold potential for >40% synergistic cognitive slowing. Oral tau inhibitors offer bioavailability advantages and improved patient convenience. The DIAN-TU platform trial is evaluating combination anti-tau and anti-amyloid therapies in familial AD at the preclinical stage. Neuroinflammation-targeting agents—directed against microglial activation, the complement cascade, and inflammatory cytokines—are under investigation as downstream neuroprotective strategies. Future treatment paradigms will likely resemble oncological or HIV treatment models, combining agents addressing distinct pathophysiological mechanisms for more substantial disease modification than monotherapy can achieve.

Conclusion

The evidence presented in this review supports a fundamental paradigm shift in dementia care: from passive management of an inevitable decline to active, sustained, multimodal intervention designed to delay cognitive deterioration across the Alzheimer's disease continuum. The key advances of this era include: proactive biomarker-driven prevention in preclinical and MCI stages; established DMT efficacy with lecanemab and donanemab achieving 27–35% cognitive slowing; cognitive rehabilitation equivalence to DMTs in resource-limited settings (~6-month delay achievable without pharmacotherapy); combined approaches achieving synergistic preservation (10–12+ month delay potential); and primary prevention trials poised to extend DMTs to asymptomatic populations.

Rehabilitation is not physical therapy for the brain, nor occupational therapy for cognition. It is a targeted, mechanistically informed biological intervention that actively engages neural plasticity to preserve function. When optimally delivered—personalised, intensive, sustained, and technology-enhanced—we are practising neuroscience, not supportive care. The tools exist. The neuroscience supports their use. The challenge ahead is implementation: changes in clinical practice, reimbursement models, training programmes, and how we conceptualise cognitive disorders. Now we must ensure these advances reach the patients who need them most.

References

Belloy, M. E., Napolioni, V., & Greicius, M. D. (2024). A quarter century of APOE and Alzheimer's disease: Progress to date and the path forward. *Nature Medicine, 30*(1), 15–26.

Buckley, R. F., Mormino, E. C., Rabin, J. S., et al. (2019). Sex differences in the association of global amyloid and regional tau distributions with cognitive decline. *JAMA Neurology, 76*(5), 542–551.

Clare, L., Kudlicka, A., Oyebode, J. R., et al. (2019). Individual goal-oriented cognitive rehabilitation to improve everyday functioning for people with early-stage dementia: A multicentre randomised controlled trial (the GREAT trial). *International Journal of Geriatric Psychiatry, 34*(5), 709–721.

Dichgans, M., & Leys, D. (2017). Vascular cognitive impairment. *Circulation Research, 120*(3), 573–591.

Dubois, B., Hampel, H., Feldman, H. H., et al. (2016). Preclinical Alzheimer's disease: Definition, natural history, and diagnostic criteria. *Alzheimer's & Dementia, 12*(3), 292–323.

GBD 2019 Dementia Forecasting Collaborators. (2022). Estimation of the global prevalence of dementia in 2019 and forecasted prevalence in 2050. *The Lancet Public Health, 7*(2), e105–e125.

Hamelin, L., Lagarde, J., Dorothée, G., et al. (2016). Early and protective microglial activation in Alzheimer's disease: A prospective study using 18F-DPA-714 PET imaging. *Brain, 139*(4), 1252–1264.

Hansson, O., Edelmayer, R. M., Boxer, A. L., et al. (2022). The Alzheimer's Association's appropriate use recommendations for blood biomarkers in Alzheimer's disease. *Alzheimer's & Dementia, 18*(12), 2669–2686.

Jack, C. R., Jr., Bennett, D. A., Blennow, K., et al. (2018). NIA-AA research framework: Toward a biological definition of Alzheimer's disease. *Alzheimer's & Dementia, 14*(4), 535–562.

Jack, C. R., Jr., Knopman, D. S., Jagust, W. J., et al. (2010). Hypothetical model of dynamic biomarkers of the Alzheimer's pathological cascade. *Lancet Neurology, 9*(1), 119–128.

Koch, G., Casula, E. P., Bonsì, P., et al. (2022). Precuneus magnetic stimulation for Alzheimer's disease: A randomised, sham-controlled trial. *Brain, 145*(11), 3776–3786.

Lefaucheur, J.-P., Aleman, A., Baeken, C., et al. (2020). Evidence-based guidelines on the therapeutic use of repetitive transcranial magnetic stimulation (rTMS): An update (2014–2018). *Clinical Neurophysiology, 131*(2), 474–528.

Leng, Y., McEvoy, C. T., Allen, I. E., & Yaffe, K. (2017). Association of sleep-disordered breathing with cognitive function and risk of cognitive impairment: A systematic review and meta-analysis. *JAMA Neurology, 74*(10), 1237–1245.

Livingston, G., Huntley, J., Sommerlad, A., et al. (2020). Dementia prevention, intervention, and care: 2020 report of the Lancet Commission. *The Lancet, 396*(10248), 413–446.

Mane, R., Chouhan, T., & Guan, C. (2020). BCI for stroke rehabilitation: Motor and beyond. *Journal of Neural Engineering, 17*(4), 041001.

Northey, J. M., Cherbuin, N., Pumpa, K. L., Smee, D. J., & Rattray, B. (2018). Exercise interventions for cognitive function in adults older than 50: A systematic review with meta-analysis. *British Journal of Sports Medicine, 52*(3), 154–160.

Palmqvist, S., Janelidze, S., Quiroz, Y. T., et al. (2020). Discriminative accuracy of plasma phospho-tau217 for Alzheimer's disease vs other neurodegenerative disorders. *JAMA, 324*(8), 772–781.

Sims, J. R., Zimmer, J. A., Evans, C. D., et al. (2023). Donanemab in early symptomatic Alzheimer disease: The TRAILBLAZER-ALZ 2 randomized clinical trial. *JAMA, 330*(6), 512–527.

Sperling, R. A., Aisen, P. S., Beckett, L. A., et al. (2011). Toward defining the preclinical stages of Alzheimer's disease: Recommendations from the National Institute on Aging–Alzheimer's Association workgroups. *Alzheimer's & Dementia, 7*(3), 280–292.

Stern, Y. (2012). Cognitive reserve in ageing and Alzheimer's disease. *Lancet Neurology, 11*(11), 1006–1012.

Tariot, P. N., Farlow, M. R., Grossberg, G. T., et al. (2004). Memantine treatment in patients with moderate to severe Alzheimer's disease already receiving donepezil. *JAMA, 291*(3), 317–324.

van Dyck, C. H., Swanson, C. J., Aisen, P., et al. (2023). Lecanemab in early Alzheimer's disease. *New England Journal of Medicine, 388*(1), 9–21.

Vemuri, P., Lesnick, T. G., Przybelski, S. A., et al. (2012). Effect of lifestyle activities on Alzheimer's disease biomarkers and cognition. *Annals of Neurology, 72*(5), 730–738.

Wang, J. X., Rogers, L. M., Gross, E. Z., et al. (2014). Targeted enhancement of cortical–hippocampal brain networks and associative memory. *Science, 345*(6200), 1054–1057.

Williamson, J. D., Pajewski, N. M., Auchus, A. P., et al. (2019). Effect of intensive vs standard blood pressure control on probable dementia: A randomized clinical trial. *JAMA, 321*(6), 553–561.

Xie, L., Kang, H., Xu, Q., et al. (2013). Sleep drives metabolite clearance from the adult brain. *Science, 342*(6156), 373–377.

Cellular Cognition in Neuro-Immune Systems

Dr. Chandra Shekhar Mayanil

Retired Principal Scientist, Lurie Children's Hospital, Chicago, IL, USA

Abstract

Cellular cognition is the capacity of individual cells to sense, integrate, and respond to complex information. It offers a powerful lens for understanding the dynamic interplay between the nervous and immune systems. This talk explores neuro-immunology through the framework of cognition at the cellular scale, highlighting how neurons, glia, and immune cells engage in reciprocal information processing that shapes health, disease, and adaptation. Immune cells exhibit hallmark cognitive functions: recognition through receptor diversity, decision-making via integrated signaling thresholds, learning through epigenetic reprogramming, and memory encoded in long-lived cellular states. In parallel, neural circuits modulate immune behavior through neurotransmitters, neuropeptides, and autonomic pathways, forming a bidirectional communication network that resembles a distributed cognitive system.

We will examine key mechanisms, including the immunological synapse, neuroimmune synapses, metabolic-epigenetic coupling, and trained immunity, that reveal how cells compute context and enact coordinated responses. Case studies from infection, autoimmunity, neurodegeneration, and immunotherapy will illustrate how cellular cognition shapes clinical outcomes. By reframing neuroimmunology through this integrative perspective, the talk argues for a unified model in which cognition is not restricted to the brain but emerges across interacting cellular networks. This conceptual shift opens new avenues for therapeutic design, from precision immunomodulation to neuro-immune circuit engineering.

Introduction: What is Cellular Cognition

Cells are active participants in biological processes. The concept of Cellular Cognition emerged following comprehensive studies of receptor-mediated cell signaling, first analyzed in isolation and subsequently reconstituted to mimic conditions within a living organism. These investigations yielded conflicting results. Therefore, a systems Biology approach was adopted, sequentially integrating additional systems to better understand cellular behavior under these circumstances. Cells perceive their surroundings, integrate multiple signals, and execute decision-making processes analogous to cognitive functions (Levin 2019; Levin 2021; Lyon 2006).

Cognitive Functions of Neural Cells

Cognitive functions of neural cells are governed by (1) Plasticity, (2) Neurotransmission, and (3) Adaptation to a changing environment.

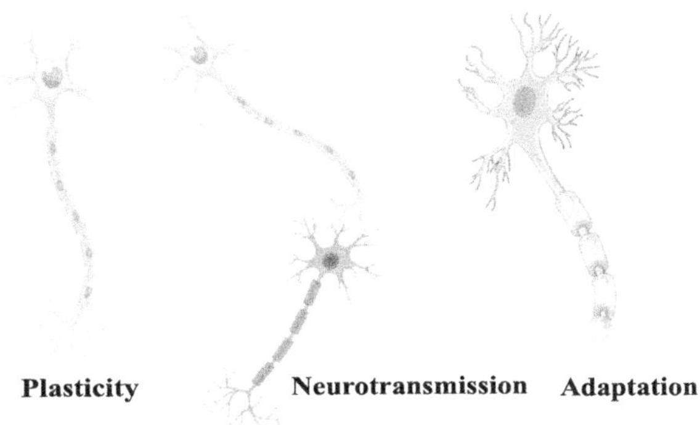

Plasticity

Neurons continuously modify both their structure and the strength of their connections. Dendritic spines can alter their shape and size, while synaptic strength may either strengthen or weaken. These processes are fundamental to learning and the formation of memories (Bliss and Collingridge 1993; Dantzer 2018).

Neurotransmission

Neuronal communication occurs through the exchange of chemical messengers. Neurotransmitter vesicles release molecules such as dopamine, glutamate, and GABA, which traverse the synaptic cleft to bind with receptors on adjacent neurons. This process facilitates swift and specific signaling throughout neural circuits (Dantzer 2018).

Adaptation

Neurons reshape their networks in response to experiences. When a stimulus repeats, it encourages the formation of new synaptic links. Their firing patterns shift to represent what has been learned, capturing the principle that neurons firing in unison become interconnected. This process forms the core of experience-dependent plasticity and cognitive flexibility. Together, these functions empower neurons to detect, interpret, and acquire knowledge, underpinning the cellular basis of cognition.

Cognitive Functions of Immune Cells

Immune cells recognize (1) patterns, (2) integrate multiple signals, and (3) retain memory through epigenetic modifications.

Pattern Recognition

Immune cells such as macrophages and dendritic cells recognize Pathogen-Associated Molecular Patterns (PAMPs) through surface receptors like toll-like receptors (TLRs), NOD-like receptors (NLRs), and RIG-I-like receptors (RLRs). These receptors interact with microbial components, including lipopolysaccharides (LPS), flagellin, and double-stranded RNA (dsRNA), which activate downstream signaling pathways (such as NF-κB and IRF) to initiate immune responses (Medzhitov 2007).

Pattern Recognition

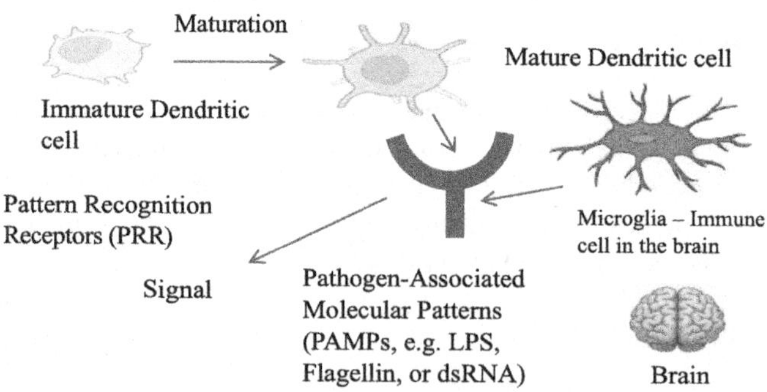

Signal Integration

Immune cells process a wide array of signals, such as cytokines, antigens, costimulatory molecules, and environmental cues, by channeling them through key intracellular pathways. The MAPK, NF-κB, and calcium signaling routes intersect within the cell. This integration shapes how the cell responds, leading to outcomes like activation, tolerance, or differentiation. In many ways, this mechanism resembles the computational strategies employed by neurons (Netea 2013).

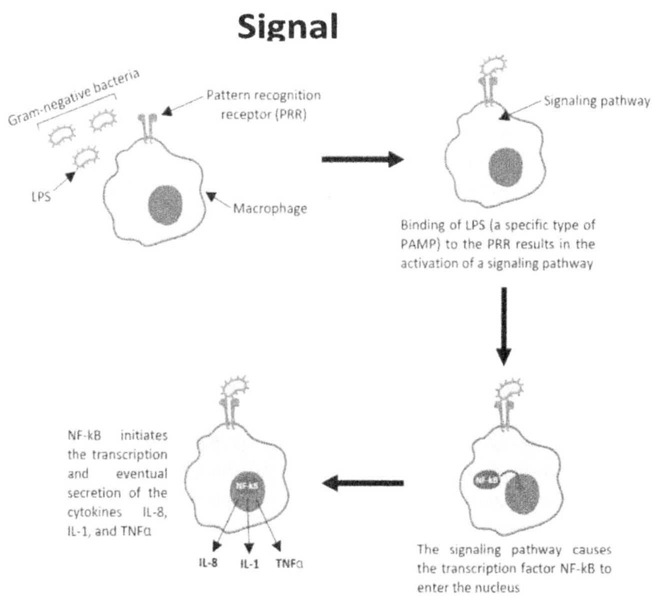

[The above infographic is from Wikimedia Commons]

Epigenetic Memory

Immune cells exhibit long-term memory of previous exposures through stable epigenetic modifications. Both DNA methylation (Me) and histone acetylation (Ac) facilitate the regulation of gene accessibility. Enzymes, including DNMTs, HDACs, and HATs, are responsible for maintaining transcriptional states. These mechanisms contribute to both trained immunity and adaptive recall in T and B lymphocytes (Netea et al., 2016).

Epigenetic Memory in Immune Cells

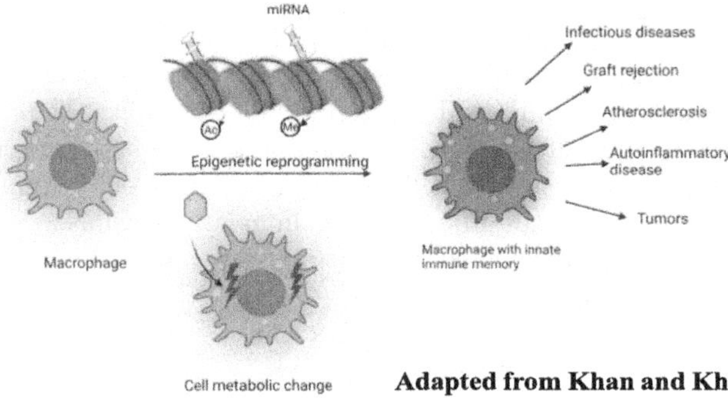

Adapted from Khan and Khan (2022)

Neuro-Immune Communication Pathways

Communication between the nervous and immune systems occurs through chemical messengers such as (1) Cytokines and neurotransmitters, (2) Vagus nerve signaling, and (3) direct glial-immune cell interactions.

Cytokines and Neurotransmitters

There is bidirectional communication between the nervous and immune systems through neurotransmitters released by neurons to influence immune cell behavior, and cytokines secreted by immune cells to modulate neuronal activity. Additionally, cellular interactions between neurons and immune cells enable precise signaling and capture the dynamic crosstalk that underlies neuroimmune regulation, which is essential for homeostasis, inflammation, and healing.

Cytokines and

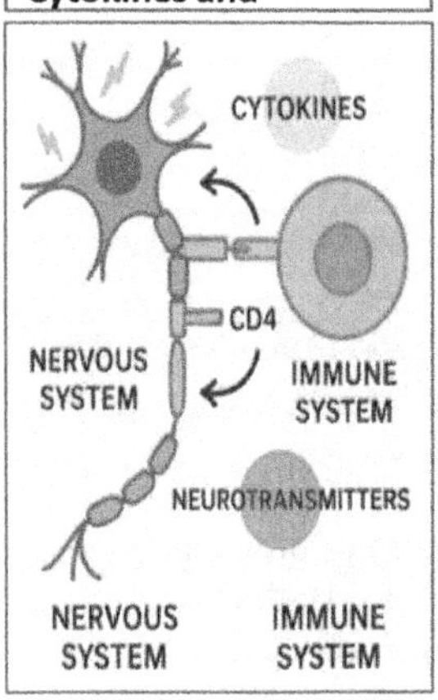

a) *Nervous System communicating with the immune system.* The electrical activity traveling down the axon of the neurons causes the release of chemical messengers, such as Norepinephrine, Acetylcholine, Dopamine, and Neuropeptides. **These molecules bind to receptors on immune cells and can** increase or decrease immune activation, shift cytokine profiles, influence T-cell differentiation, and regulate inflammation. **This is the basis of neuro-immune modulation.**

b) *Immune System communicating with the nervous system.* The immune cell releases cytokines, such as IL-1β and TNF-α. IL-6 and IFN-γ can act on neurons and glial cells to (1) alter synaptic activity, (2) change neural excitability, (3) influence mood, cognition, and fatigue, and (4) trigger sickness behavior. **This explains why inflammation affects brain function.**

c) *CD4 as a Molecular Bridge.* The CD4 represents a key immune receptor found on helper T cells. $CD4^+$ T cells influence (1) Microglial activation, (2) Neurogenesis, (3) Synaptic pruning, and (4) Long-term neural plasticity. CD4 is shown as a symbolic "bridge" between the two systems, emphasizing that immune cell identity and activation state shape neural outcomes.

Vagal Nerve Activation

Vagus nerve activation acts as an essential link connecting the nervous system to the immune system (Craig 2003; Porges 2007). Starting at the brainstem, the vagus nerve sends signals to organs such as the spleen and gut, where immune cells like macrophages and T cells detect these signals through receptors including α7nAChR (alpha-7 nicotinic acetylcholine receptor).

Vagus Nerve Signaling

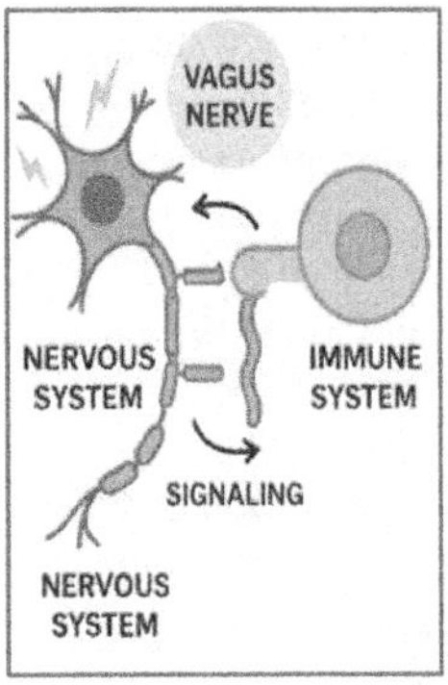

Bidirectional signaling: **Vagus nerve afferent fibers** detect inflammation and relay signals to the brain. **Efferent fibers,** on the other hand, release **acetylcholine**, which suppresses pro-inflammatory cytokines like **TNF-α** and **IL-1β**. This pathway is central to the **inflammatory reflex**, a neural circuit that regulates immune responses and maintains homeostasis (Tracey 2002).

This infographic shows how neurons and immune cells form synapse-like structures to exchange information. It highlights bidirectional communication via neurotransmitters, cytokines, and receptor-ligand interactions. On the right: an immune cell (e.g., microglia or a T cell) forming a structured junction with the neuron. The synaptic cleft between them shows arrows for two-way signaling.

Glial-Immune Interaction

Glial-Immune Interaction

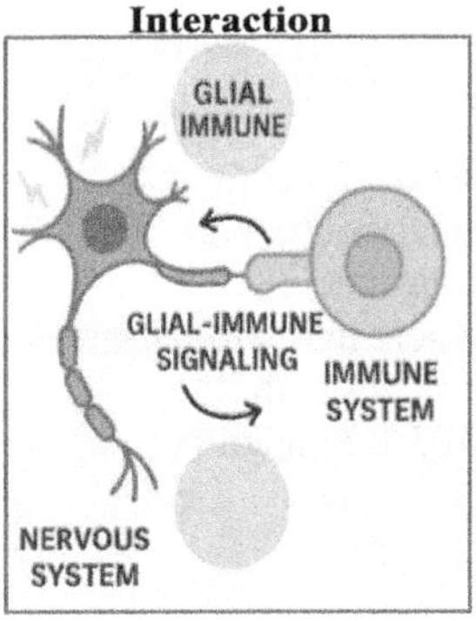

A Two-Way Regulatory Loop: The nervous system regulates immunity **via** neurotransmitters, neuropeptides, and autonomic pathways. The immune system regulates neural function **via** cytokines, chemokines, and immune-cell–glial interactions. **This loop is central to** stress responses, infection and sickness behavior, autoimmunity, neurodegeneration, pain, and fatigue syndromes. Neurons and immune cells communicate through shared molecular languages, cytokines, neurotransmitters, and receptor signaling. This communication shapes plasticity, inflammation, behavior, and overall health. This figure shows how cells, especially neurons and immune cells, perform **signal integration** and make **context-sensitive decisions**.

Scientific Explanations

External signals, such as neurotransmitters, arrive at the immune cells; cytokines arrive at the neural cells. These are processed in a central hub, the cell's **signal integration machinery**.

The cell then chooses one of several **decision outcomes**: (1) **Activation** (e.g., immune response, neural firing), (2) **Tolerance** (e.g., immune suppression, neural inhibition), (3) **Memory** (e.g., epigenetic imprinting, synaptic plasticity). Signals such as **cytokines**, **metabolic inputs**, and **environmental stressors** influence chromatin. Epigenetic modifications (e.g., methylation, acetylation) alter gene expression. These marks persist, forming **epigenetic memory**. The cell "remembers" past exposures and responds accordingly.

Yogic Neuroscience Interpretation

The **signal integration hub** mirrors ***Chitta***, the subtle field of consciousness. Incoming signals are like ***Vrittis***, fluctuations in awareness. When signals are **coherent**, *Chitta* becomes ***Prasanna*** (clear, calm). The **decision outcomes** reflect **karmic tendencies**: Activation equates to *rajas* (action), Tolerance equates to *sattva* (equanimity), and Memory is equivalent to *samskāra* (imprint).

Epigenetic marks are **cellular *samskāras***, stored tendencies. Just as *samskāras* shape mental behavior, epigenetic memory shapes cellular behavior. Yogic purification (e.g., mantra, breath, *dhyāna*) may help **repattern these tendencies**, both mentally and biologically. This model suggests that **cellular decision-making is a microcosmic reflection of mental regulation**, and that Yogic practices may help stabilize cellular cognition.

Cellular Cognition and Yogic Concepts

Cells integrate signals, neurotransmitters, cytokines, hormone through dynamic feedback loops.
When signals are **coherent**, cells stabilize their internal states (like ***chitta***, the calm field of consciousness).
When signals are **incoherent**, cells fluctuate, activating stress pathways, inflammatory cascades, or erratic firing (like **vrittis** or ripples in consciousness).

Cellular Cognition and Yogic Concepts

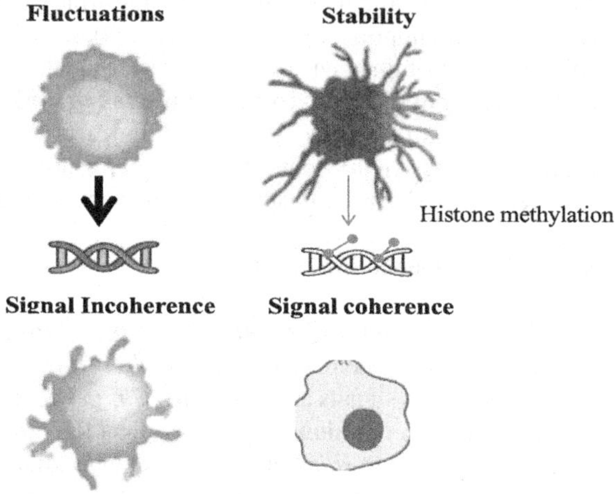

Cells fluctuate (forming Vrittis) or stabilize (like Chitta) based on signal incoherence or signal coherence

[This infographic illustrates how cellular cognition mirrors the yogic concepts of chitta and vritti]

Yogic Mind: Spiritual Interpretation

Chitta is the subtle field of awareness. **Vrittis** are its fluctuations, thoughts, emotions, and impressions. Yoga aims to regulate vrittis and restore chitta's stillness through practices like **pranayama**, **dhyana**, and **pratyahara**.

Bridging Insight

Just as neurons and immune cells stabilize or fluctuate based on **signal coherence**, the yogic mind stabilizes or ripples based on **mental impressions and pranic flow**.

This parallel suggests that **yogic practices** may help regulate cellular Cognition states, supporting clarity, resilience, and healing.

Cellular Cognition & Chitta-Vritti Regulation

Prana corresponds to metabolic and epigenetic dynamics by acting as a subtle regulatory force that modulates cellular energy states, redox balance, and gene expression. Yogic practices like pranayama, dhyāna, and mantra actively shape these flows, enhancing coherence, resilience, and long-term cellular memory (Brown and Gerbarg, 2005).

PRANA, METABOLIC &
EPIGENETIC DYNAMICS

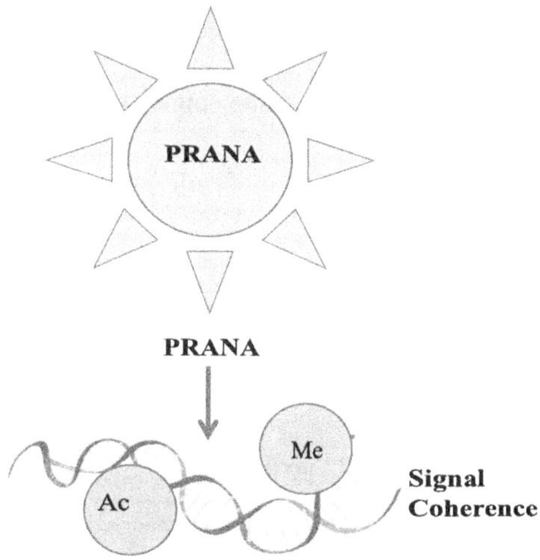

Metabolic and Epigenetic Dynamics

Scientific Layer: Prana as Biochemical Information

- Acetyl-CoA, NAD^+, and lactate are metabolic intermediates that influence epigenetic enzymes:
- HATs (histone acetyl transferases) use acetyl-CoA to open chromatin.
- DNMTs (DNA methyltransferases) use SAM to silence genes.
- α-KG supports demethylation via TET enzymes. TET enzymes, or Ten-Eleven Translocation (TET), are a family of dioxygenases (TET1, TET2, TET3) that play a critical role in epigenetics by oxidizing 5-methylcytosine to 5-hydroxymethylcytosine, which promotes DNA demethylation. They act as

"erasers" of DNA methylation, regulating gene expression, cell differentiation, and maintaining genome stability.

The diagram above shows:

- Prana as a central energetic force.
- Its influence on metabolic molecules like acetyl-CoA and NAD^+.
- These molecules shape epigenetic marks (Me, Ac) on DNA.
- Yogic practices modulate pranic flow, thereby influencing cellular memory and resilience.

These molecules act as informational substrates, encoding cellular experience into gene expression.

Modulation Through Practice: Yogic Interpretations

Pranic Concept	**Biological Correlate**
Prana (vital energy)	Metabolic flux: ATP production, redox state, oxygen utilization
Samskāras (impressions)	Epigenetic marks: DNA methylation (Me), histone acetylation (Ac)
Pranic flow	Bioelectrical and biochemical signaling, vagal tone, Heart Rate Variability, NAD^+/NADH ratio
Disturbed prana	Inflammatory cytokines, oxidative stress, and mitochondrial dysfunction

These practices entrain pranic rhythms, which in turn reshape metabolic and epigenetic landscapes, leading to long-term changes in cognition, immunity, and emotional regulation (Bernardi et al 2001; Kaliman et al, 2014; Streeter et al, 2012; Thayer and Lane, 2000).

Yogic Practice	**Physiological Effect**
Pranayama	Alters CO_2/O_2 ratio, causes a shift in redox state, which modulates epigenetic enzymes
Dhyāna (meditation)	Enhances vagal tone, reduces cortisol, and stabilizes metabolic and immune rhythms
Mantra & Mudra	Entrains breath and attention, synchronizes neural oscillations, and improves Heart Rate Variability
Bandhas & Kriyas	Stimulate cerebrospinal and lymphatic flow, detoxify, and reset cellular signaling

Definition of Health, Survival, and Evolution Through the Lens of Integrated Cellular Intelligence

When neural cell intelligence combines immune intelligence and pranic intelligence, it can create super metacognitive intelligence at the cellular level, a synthesis that connects neurobiology, immunology, and bioenergetics. This forms what is essentially the **biological "Internet of Things"**: a network in which every node functions not just as a passive receiver, but as an active, self-aware processor.

To understand how these three create "Super Metacognitive Intelligence," we have to look at how they exchange information and regulate one another.

Neural Cell Intelligence: The High-Speed Processor

Neural intelligence isn't just about "thinking"; at a cellular level, it's about **synaptic plasticity** and rapid signal transduction. Neurons use electrochemical gradients to process information at millisecond speeds.

a) **The Metacognitive Role:** It provides "top-down" architecture. It maps the external world and internal states, creating a predictive model of what the body needs next.
b) **Key Mechanism:** The release of neurotransmitters that don't just talk to other neurons but also signal immune cells to "wake up" or "calm down."

Immune Intelligence: The Decentralized Memory

The immune system is often called the "mobile brain." It possesses a sophisticated memory (T and B cells) and an incredible ability to distinguish "self" from "non-self."

a) **The Metacognitive Role:** It acts as the body's "quality control." While the brain processes abstract data, the immune system processes **molecular reality**. It learns from every encounter with a pathogen, storing that data in a decentralized network that spans the entire body.
b) **The Bridge:** Immune cells produce cytokines, small signaling proteins, that cross the blood-brain barrier to influence neural behavior, mood, and even cognition.

Pranic Intelligence: The Bioenergetic Engine

While "Prana" is a traditional term for life force, in a biological context, we can view this as **Bioenergetic Intelligence**, the flow of protons and electrons within the mitochondria and the subtle electromagnetic fields generated by cellular activity (Brown and Gerbarg, 2005; Picard and McEwen, 2018; Wallace, 2005, 2010).

a) **The Metacognitive Role:** This is the "power management" layer. No intelligence can function without ATP (adenosine triphosphate). Pranic intelligence regulates the **efficiency of energy conversion**. If a cell's bioenergetic field is disrupted, the neural and immune responses falter.
b) **The subtle flow:** Research into "biofields" suggests that cells communicate via biophotons (weak light emissions), coordinating complex activities faster than chemical signals alone could allow.

The Synthesis: Super Metacognitive Intelligence

When these three systems converge, they create a **recursive feedback loop**—this is where "metacognition" (thinking about thinking or sensing the sensing) happens at a cellular level.

How they integrate:

a) **The Neuro-Immune Axis:** The brain senses a threat (Neural) and triggers a stress response. The immune system receives this signal and prepares for inflammation (Immune system).
b) **The Bioenergetic Governor:** The mitochondria (Pranic) determine if there is enough "fuel" to sustain this response. If energy is low, the system "throttles" the neural activity to prevent burnout (Picard & McEwen, 2018; Wallace, 2005, 2010).
c) **Self-Correction:** The immune system "reports back" to the brain via the Vagus nerve, telling the brain whether its initial assessment of the threat was correct.

This creates Super-**Intelligence** because the body isn't just reacting; it is **evaluating its own reactions** and adjusting its energetic, chemical, and electrical outputs in real-time.

Neural intelligence offers *logic*, immune intelligence supplies *context/memory*, and pranic intelligence delivers *vitality*. Combined, these elements create a self-aware biological system supporting health, survival, and evolution. The Vagus Nerve is key in enabling this Super Metacognitive State.

Understanding the Vagus nerve means grasping the concept of your body's **biological motherboard**. As the longest cranial nerve, it extends from the brainstem down to your lowest organs, serving as a two-way superhighway that connects and coordinates the three forms of intelligence mentioned above.

The Vagus Nerve Facilitates a "Super Metacognitive" State

The Interoceptive Sensor (Neural ↔ Immune)

About **80% of the fibers** in the Vagus nerve are afferent, meaning they carry information *from* the body *to* the brain. It is the primary organ of **Interoception**, the sense of the internal state of the body.

a) **The Metacognitive Loop:** The Vagus nerve "listens" to the cytokines produced by immune cells in the gut and lungs. It then tells the brain, "The environment is inflamed; we need to shift into a defensive mode."
b) **Action:** The brain then adjusts your behavior (sickness behavior like lethargy) to conserve energy for healing.

The Cholinergic Anti-Inflammatory Pathway (Immune ↔ Neural)

This is where the Vagus nerve acts as a "remote control" for the immune system. When the brain detects excessive inflammation, it sends a signal back down the Vagus nerve.

a) **The Mechanism**: The nerve releases acetylcholine near immune cells (like macrophages).
b) **The Result**: Acetylcholine binds to receptors on these cells and effectively "turns off" the production of pro-inflammatory chemicals. This is a master-level metacognitive feat: the neural system is literally editing the immune system's response to prevent self-damage.

Heart Rate Variability (Pranic ↔ Neural)

In bioenergetics (Pranic Intelligence), the Vagus nerve is the master regulator of **Heart Rate Variability (HRV)**. HRV is the measure of the variation in time between each heartbeat.

a) **High HRV** indicates a flexible, resilient system where Pranic energy is flowing efficiently. The Vagus nerve is "toned" and can quickly switch the body from stress to recovery.
b) **Low HRV** suggests the system is "stuck" in a stress loop, draining the body's energetic battery (ATP).
c) **Synergy** by using "*Pranic*" techniques like deep, rhythmic breathing, you manually stimulate the Vagus nerve. This sends a signal to the brain that the environment is safe, which in turn lowers inflammation and optimizes energy distribution.

The Intelligence Integration Table

Intelligence Type	Vagal Function	Metacognitive Outcome
Neural	Fast-track signaling	**Situational Awareness:** Real-time mapping of internal organ status.
Immune	Cytokine monitoring	**System Integrity:** Prevents the immune system from overreacting (Autoimmunity).
Pranic	HRV Regulation	**Energy Efficiency:** Optimizes the "burn rate" of cellular fuel based on environmental demand.

The "Super" Intelligence Result

When these three components synchronize through the Vagus nerve, the body attains **Hemodynamic Flow**. This state extends beyond mere "balance" (homeostasis); it represents an advanced, adaptive response. The system is enhanced, enabling prediction of potential threats, efficient management of energy resources, and simultaneous self-repair.

The significance of "gut feelings" is underscored by the Vagus nerve's role in transmitting intricate immune and microbial information to the brain before conscious processing of the situation.

In Summary, Neural circuits regulate immunity; Immune signals shape plasticity; and Pranic rhythms coordinate both. **Coherence among them supports health.**

References

Bernardi, L., Sleight, P., Bandinelli, G., et al. (2001). Effect of rosary prayer and yoga mantras on autonomic cardiovascular rhythms: Comparative study. *BMJ, 323*(7327), 1446–1449.

Bliss, T. V. P., & Collingridge, G. L. (1993). A synaptic model of memory: Long-term potentiation in the hippocampus. *Nature, 361*(6407), 31–39.

Brown, R. P., & Gerbarg, P. L. (2005). Sudarshan Kriya yogic breathing in the treatment of stress, anxiety, and depression: Part I—Neurophysiologic model. *Journal of Alternative and Complementary Medicine, 11*(1), 189–201.

Craig, A. D. (2003). Interoception: The sense of the physiological condition of the body. *Current Opinion in Neurobiology, 13*(4), 500–505.

Dantzer, R. (2018). Neuroimmune interactions: From the basics to the clinical. *Psychological Science in the Public Interest, 19*(1), 4–31.

Khan, S. U., & Khan, M. U. (2022). Advances in innate immune memory of macrophages. *Exploration of Immunology, 2*, 428–441.

Kaliman, P., et al. (2014). Rapid changes in histone deacetylases and inflammatory gene expression in expert meditators. *Psychoneuroendocrinology, 40*, 153–160.

Levin, M. (2019). The computational boundary of a self: Developmental biology and memory. *Frontiers in Psychology, 10*, Article 2688.

Levin, M. (2021). Bioelectric networks: The cognitive glue of embryonic development and regeneration. *Frontiers in Cell and Developmental Biology, 9*, Article 644442.

Lyon, P. (2006). The cognitive cell: Bacterial behavior reconsidered. *Frontiers in Ecology and the Environment, 4*(7), 345–353.

Medzhitov, R. (2007). Recognition of microorganisms and activation of the immune response. *Nature, 449*(7164), 819–826.

Netea, M. G. (2013). Trained immunity: An antigen-independent molecular memory in innate cells. *Trends in Immunology, 34*(5), 204–214.

Netea, M. G., Joosten, L. A. B., et al. (2016). Trained immunity: A program of innate immune memory in health and disease. *Science, 352*(6284), Article aaf1098.

Picard, M., & McEwen, B. S. (2018). Psychological stress and mitochondria: A conceptual framework. *Psychosomatic Medicine, 80*(2), 126–140.

Porges, S. W. (2007). The polyvagal perspective. *Biological Psychology, 74*(2), 116–143.

Streeter, C. C., et al. (2012). Effects of yoga on the autonomic nervous system, gamma-aminobutyric acid, and allostasis in epilepsy, depression, and posttraumatic stress disorder. *Medical Hypotheses, 78*(5), 571–579.

Thayer, J. F., & Lane, R. D. (2000). A model of neurovisceral integration in emotion regulation and dysregulation. *Journal of Affective Disorders, 61*(3), 201–216.

Tracey, K. J. (2002). The inflammatory reflex. *Nature, 420*(6917), 853–859.

Wallace, D. C. (2005). A mitochondrial paradigm of metabolic and degenerative diseases, aging, and cancer. *Genetics, 171*(2), 383–405.

Wallace, D. C. (2010). Bioenergetics and the epigenome. *Nature Genetics, 42*(11), 916–920.

Quantum Artificial General Intelligence: Exploring the Convergence of Quantum Computing and Cognitive Architectures

Dr. Mrityunjay Guha Majumdar

Adjunct Professor, National Institute of Advanced Studies, Bengaluru, India; Chair, Quantum Information and Intelligence Systems Group, Amrita Vishwa Vidyapeetham, Delhi NCR, India

Abstract

Artificial General Intelligence (AGI) aims to achieve human-level adaptability and cognition, yet classical computational approaches remain constrained by their underlying physics. Quantum mechanics offers a radically different foundation for intelligence, one shaped by principles such as non-locality, contextuality, and the no-cloning theorem. These quantum features redefine how information can be represented, learned, and generalized, suggesting that a Quantum AGI (QAGI) may differ fundamentally from classical AGI in both capability and ontology. This talk explores how results such as Bell's theorem and the Kochen–Specker theorem introduce new forms of reasoning and correlation inaccessible to classical systems, while coherence and decoherence dynamics set practical bounds on learning and memory. We highlight emerging frameworks—quantum neural networks, quantum reinforcement learning, and quantum reservoir computing-as steps toward adaptive, physically grounded intelligence. Quantum AGI represents not simply a faster form of computation but a new mode of cognition shaped by the structure of quantum reality, demanding re-examination of intelligence, embodiment, and agency from first physical principles.

Keywords: Quantum Artificial General Intelligence, Quantum Machine Learning, Cognitive Architectures, Quantum Cognition, Hybrid Quantum-Classical Systems, Consciousness Studies

Introduction

The central question driving this investigation is whether quantum principles can facilitate the development of Artificial General Intelligence. This inquiry emerges at the intersection of two transformative scientific domains: the quest for human-level artificial intelligence and the revolution in quantum information science. As classical computing approaches encounter fundamental limitations in replicating the full spectrum of human cognition, quantum mechanics offers novel computational paradigms that may address these constraints. Artificial General Intelligence represents a qualitative departure from contemporary narrow AI systems. AGI is defined as a computational system capable of achieving robust, autonomous, and transferable competence across the full range of cognitive tasks accessible to humans, without task-specific redesign or retraining. This definition encompasses several essential capabilities. AGI must construct and apply unified internal representations that support reasoning across heterogeneous tasks and environments, enabling knowledge transfer between domains that share underlying structures despite surface-level differences.

The system must recombine previously acquired abstractions and skills to solve novel problems with minimal additional data, demonstrating the human capacity for flexible knowledge application. AGI requires calibrated uncertainty estimation and adaptive strategy selection based on its own epistemic state, maintaining awareness of what it knows and the reliability of its knowledge. Rather than relying solely on surface-level statistical correlations, AGI must capture invariant causal structure in the world, forming representations that support intervention and counterfactual reasoning. The system must plan over long horizons and adapt behavior under uncertainty to optimize temporally extended objectives, maintaining coherent goal structures across diverse contexts. It is crucial to distinguish AGI from large language models, as next-token prediction alone does not constitute unified agency, causal world modeling, or autonomous general intelligence. The scale and sophistication of current AI systems should not be conflated with genuine general intelligence.

Current AI systems, despite impressive advances, exhibit fundamental limitations that preclude AGI status. Contemporary systems optimize task-specific loss functions under constrained training distributions rather than optimizing for domain-general competence or unified cognitive performance. Each new task typically requires architectural modifications or retraining. Modern deep learning architectures require large-scale labeled or self-supervised datasets to approximate target functions, lacking the sample-efficient structure learning and rich inductive biases characteristic of human cognition. Humans can learn from a handful of examples, whereas machines require millions. Performance degrades significantly under distributional shift because learned representations encode statistical regularities of the training manifold rather than invariant causal structure. This brittleness manifests in adversarial vulnerability and failure in novel contexts. Current systems are typically modular, task-fragmented, or pipeline-based, without integrated architectures that coherently unify perception, memory, reasoning, planning, and self-modeling.

The seamless integration characteristic of human cognition remains elusive. State-of-the-art models require orders of magnitude more compute and energy than biological intelligence to achieve comparable performance on bounded tasks, reflecting limited algorithmic efficiency and poor structural priors. The human brain operates on approximately 20 watts while contemporary AI systems consume megawatt-scale resources. The scale of neural complexity in even small brain regions is humbling. Recent mapping of just one cubic millimeter of human brain revealed approximately 57,000 cells and 150 million synaptic connections, consuming 1.4 petabytes of storage. The entire human brain contains roughly 86 billion neurons with an estimated 100 trillion synapses—more connections than stars in the observable universe. Yet even complete connectome mapping would not necessarily illuminate how consciousness and intelligence emerge from these connections.

Compounding these challenges, up to 30% of the power used to train modern AI models may be wasted. Data centers could account for 1.2% of global carbon emissions by 2027, and AGI development under classical computing paradigms would exponentially increase this figure. The problem becomes even more complex when considering how to measure reasoning—a core intelligence component. Current benchmarks test specific capabilities but often fall prey to Goodhart's Law: once a metric becomes a target, it ceases to be a good metric. Models can be optimized for specific benchmarks without developing general reasoning capacities, raising fundamental questions about what would constitute proof of reasoning in artificial systems.

Bio-Inspired Cognitive Architectures

Bio-inspired cognitive architectures formalize cognition as integrated control systems coupling perception, memory, learning, decision-making, and action within unified dynamical frameworks. These architectures attempt to capture the functional organization of biological intelligence while remaining computationally tractable. *Perception* is modeled as hierarchical inference over structured representations, transforming multimodal sensory inputs into latent symbolic or sub-symbolic states. Mechanisms include production rules in ACT-R (Adaptive Control of Thought—Rational), chunking in SOAR (State, Operator, and Result), and predictive coding hierarchies in Predictive Processing frameworks. *Memory* is typically partitioned into declarative and procedural subsystems with explicit mechanisms for encoding, retrieval, activation decay, and working memory constraints. This enables temporally extended reasoning and state persistence while respecting the limited capacity characteristic of human working memory. *Learning* is implemented through rule induction, reinforcement learning, Bayesian updating, or error-driven prediction minimization, allowing incremental refinement of internal representations and action policies without catastrophic forgetting of previously acquired knowledge. *Decision-making* is formalized as goal-directed selection over competing productions, operators, or policies, mediated by utility estimation, conflict resolution mechanisms, or global broadcasting dynamics inspired by Global Workspace Theory. *Action* emerges from closed-loop perception-cognition-motor cycles where internal state updates continuously modulate behavior, exemplified by ACT-R's production compilation, SOAR's operator selection and chunking, Global Workspace's attentional broadcasting, and Predictive Processing's active inference framework.

Biological intelligence exhibits several features that distinguish it from artificial systems and may inform quantum approaches. Biological neural systems implement highly distributed, asynchronous computation across approximately 10^{11} neurons and 10^{14}–10^{15} synapses, enabling concurrent signal integration, local plasticity, and decentralized control without global clock synchronization. Neural processing is strongly state-dependent, with responses modulated by prior activity, neuromodulators, network oscillations, and long-range feedback loops, allowing dynamic contextual gating and task-dependent reconfiguration. Cortical dynamics approximate Bayesian belief updating through population coding, stochastic spiking, and predictive error minimization, supporting uncertainty-aware inference rather than deterministic symbol manipulation.

Sensory and cognitive pathways are organized into multi-scale, recurrent hierarchies that progressively abstract invariant structure while maintaining bidirectional signaling for top-down constraint integration. The human brain achieves high cognitive throughput under strict metabolic constraints via sparse coding, event-driven spiking, local plasticity rules, and analog computation, vastly outperforming current digital architectures in joules per inference. Biological systems exhibit lifelong plasticity through synaptic potentiation/depression, structural rewiring, and neuromodulatory regulation, enabling real-time adaptation without catastrophic forgetting.

Classical computing faces several fundamental barriers in addressing the full scope of cognitive tasks. Many cognitive tasks—planning, symbolic reasoning, hypothesis generation—require search over exponentially large state spaces, rendering exhaustive evaluation computationally intractable. The branching factor in real-world planning scenarios rapidly exceeds any feasible classical search.

Core cognitive functions, including optimal planning (POMDP policy optimization), constraint satisfaction, and probabilistic graphical model inference, are NP-hard or worse, implying no known polynomial-time solutions under standard complexity assumptions. Exact inference in high-dimensional Bayesian networks requires marginalization over exponentially many latent variables, making precise posterior computation computationally prohibitive except in restricted graph topologies. Cognitive environments involve continuous, high-dimensional state representations where required sampling density or discretization grows exponentially with dimensionality, degrading learning and inference efficiency. Practical classical systems rely on variational approximations, Monte Carlo sampling, or heuristic search, introducing bias-variance trade-offs and limiting guarantees of optimality or global coherence.

Classical von Neumann architectures impose memory bandwidth and sequential processing bottlenecks that constrain real-time integration of large-scale, recurrent, and dynamically coupled representations characteristic of cognition. Quantum computing offers fundamentally different computational primitives that may address these limitations. Unlike classical bits constrained to 0 or 1, quantum bits (qubits) can exist in superposition states, enabling simultaneous representation of multiple hypotheses. Entanglement creates non-separable correlations between subsystems, supporting holistic representations impossible in classical systems. Interference allows constructive and destructive combination of computational paths, enabling quantum algorithms that outperform classical counterparts for specific tasks.

The Quantum Information Paradigm and QAGI

The first quantum revolution transformed technology through the utilization of quantum resources in transistors and optical masers. The transistor, invented in 1947 by Bardeen, Brattain, and Shockley at Bell Laboratories, demonstrated amplification of electrical signals and marked a pivotal moment in electronics. The optical maser (laser), evolving from Townes's 1954 maser demonstration and Maiman's 1960 ruby laser, revolutionized light amplification technology. These advances paved the way for efficient electronic devices and enhanced communication systems, representing the foundational layer of quantum-enabled technology.

The second quantum revolution emerged from recognizing entanglement as a computational resource. Einstein, Podolsky, and Rosen's 1935 paper first highlighted the "spooky" features of quantum mechanics, with Schrödinger coining "Verschränkung" (entanglement) to describe global states of composite systems that cannot be written as products of individual subsystem states. Bell's 1964 work formalized the EPR argument, proving that assumptions of reality, locality, and free will impose constraints (Bell inequalities) on statistical correlations that entangled quantum states violate. This opened the door for entanglement-based technologies such as *Quantum cryptography,* which uses entangled states for secure communication, *Quantum superdense coding,* which encodes two classical bits in one qubit, and *Quantum teleportation* which transfers quantum states using entanglement and classical communication. These effects have been experimentally demonstrated and extended to entanglement manipulation, broadcast, and controlled distribution.

The emerging third quantum revolution applies quantum computing to machine learning, addressing fundamental challenges in classical learning paradigms. Quantum Machine Learning (QML) modifies classical algorithms for quantum functionality, with integration models including Quantum Data Classical Model, Classical Data Quantum Model, and Quantum Data Quantum Model. Quantum Enhanced Machine Learning (QEML) analyzes classical data with quantum models, utilizing qubits and quantum operations in cross-sectional subroutines to enhance computing speed and data storage. Computationally challenging subroutines are executed on quantum devices in hybrid approaches, with the workflow typically involving system backend, state preparation, feature mapping, unitary gate operations, and quantum measurement. Key QML algorithms include *Quantum Support Vector Machines (QSVM)* for classification and regression, implemented on trapped-ion quantum computers; *Quantum Neural Networks (QNN)* with training capabilities for deep architectures; *Quantum k-Nearest Neighbour (QkNN)* models using controlled-SWAP gates for distance computation; *Quantum Random Forests* employing kernel-based approaches for improved classification. Classical hypothesis evaluation in symbolic or sequential probabilistic systems is typically serialized, with belief states represented as single maximally probable models or updated distributions at each computational step. The quantum superposition principle offers a fundamentally different approach: quantum systems can exist in linear combinations of basis states, enabling simultaneous amplitude representation over multiple hypotheses before measurement-induced collapse.

This parallels cognitive phenomena where human perception maintains multiple competing interpretations of ambiguous stimuli. Bistable figures like the Necker Cube demonstrate transient co-activation of alternative internal models before resolution. In uncertain reasoning, cognitive systems maintain graded belief distributions over mutually incompatible hypotheses, approximating probabilistic mixtures rather than committing to single discrete states. Neural population dynamics allow simultaneous activation of partially incompatible representations, with attractor competition and inhibitory dynamics resolving toward coherent interpretations over time. Analogous to quantum measurement, cognitive decision-making can be modeled as a commitment operation reducing distributed belief states into single actions or judgments, despite prior parallel representational support. The neural binding problem addresses how biological cognition integrates distributed feature representations—color, shape, motion, location—encoded across spatially separated neural populations into unified, temporally coherent precepts without explicit central indexing. Entangled quantum states, which cannot be factorized into independent subsystem states, offer a mathematical formalism for irreducible correlations not decomposable into local properties. Unified precepts require relational encoding where feature identity is conditionally bound to others, preventing combinatorial misbinding across objects. Higher-order concepts exhibit context-sensitive meaning such that composite representations (e.g., "pet fish") cannot be derived via linear combination of constituent semantic vectors, indicating non-separable semantic structure.

Cognitive reasoning is strongly context-dependent, with proposition evaluation violating classical independence assumptions and reflecting conditional co-activation patterns across semantic networks. Both biological binding and entangled systems highlight the necessity of representing joint states with intrinsic correlations rather than aggregating independently encoded components. Quantum cognition applies quantum mathematical formalism to model cognitive and decision-making phenomena. Contextual probability models explain violations of classical probability by allowing measurement to change cognitive state. Non-commutative decision processes model sequential judgments using non-commuting operators, accounting for empirically observed order effects. Superposition and interference effects capture deviations from the classical law of total probability in tasks such as conjunction and disjunction fallacies. Crucially, quantum cognition uses quantum probability theory to describe cognitive statistics without assuming the brain is a quantum computer. This distinction is essential: mathematical formalisms developed for quantum mechanics may capture cognitive regularities even if the underlying neural substrate is classical. Quantum AGI means implementing general intelligence algorithms on quantum computing hardware, using qubits and quantum operations as a computational substrate. It concerns computational architecture—not "quantum consciousness" per se—and does not assume that intelligence or awareness is inherently quantum. This distinction is critical for maintaining scientific rigor while exploring potentially transformative approaches.

A promising approach integrates classical neural networks with dedicated quantum processing cores for computationally intensive subroutines. *Classical neural modules* handle perception, feature extraction, representation learning, memory, and high-level control using conventional deep learning frameworks, leveraging their proven capabilities in these domains. *The quantum processing core* operates on qubits in Hilbert space, applying unitary transformations and entanglement to perform structured computations. Sampling tasks may leverage quantum amplitude-based methods for efficient sampling from high-dimensional distributions. Optimization routines utilize algorithms such as QAOA or variational quantum circuits for combinatorial or non-convex problems. Associative search and inference exploit quantum amplitude amplification or linear-algebraic speedups. *The decision layer* integrates measured quantum outputs with classical policies or reinforcement learning mechanisms to produce final actions, maintaining the benefits of both paradigms. Quantum associative memory exploits superposition to store and retrieve patterns in ways impossible classically. By encoding memory items in orthogonal quantum states, quantum associative memory can perform content-addressable retrieval with exponential capacity relative to classical associative memories, though practical implementations face significant decoherence challenges. Quantum Reinforcement Learning (QRL) extends classical reinforcement learning by encoding value functions, policies, or environment models into quantum states, enabling coherent evaluation over superposed state-action configurations. Accelerated policy evaluation using quantum linear-system solvers and amplitude estimation techniques can provide polynomial or quadratic speedups in estimating value functions compared to classical Monte Carlo sampling. Grover-style search enables quadratic speedup in identifying high-reward actions within large discrete action spaces. Exploration strategies can leverage amplitude amplification to preferentially increase the sampling probability of promising state-action pairs. Quantum superposition allows simultaneous evaluation of multiple policies or trajectories, potentially reducing sample complexity in environments with large combinatorial state spaces. However, QRL remains largely theoretical and experimentally constrained by noise, limited qubit counts, and data-loading overhead. Provable speedups depend on oracle assumptions and fault-tolerant quantum hardware not yet realized.

The Global Workspace Theory (GWT), an influential cognitive architecture framework, can be formalized in Hilbert-space terms: the Global Workspace is modeled as a high-dimensional Hilbert space in which distributed cognitive representations are encoded as state vectors, enabling unified yet compositional representation of multiple neural assemblies. Pre-conscious parallel processing across specialized modules corresponds to coherent superposition of basis states, allowing simultaneous representation of multiple competing cognitive contents before selection. Attentional selection is formalized analogously to quantum measurement: interaction with a selection operator projects the superposed cognitive state onto a specific subspace, yielding a single globally broadcast representation. Feature binding across distributed cortical regions is modeled through entangled states, capturing non-separable correlations between perceptual, semantic, and affective components. The "global availability" postulated by GWT corresponds to coherent state redistribution across subsystem Hilbert spaces via tensor-product structure, enabling rapid integration across functional modules. Competing cognitive representations can be described as interference patterns within the workspace, with decoherence-like processes modeling stabilization into classical-like, reportable conscious states. While empirically unverified at the physical level, the Hilbert-space formalism provides mathematically rigorous unification of superposition, selection, and relational binding, aligning conceptually with interdisciplinary themes linking cognition, computation, and quantum theory.

Major Challenges and Research Roadmap

There are a few technical challenges that we must consider when we talk of quantum AGI and approaches towards understanding cognition. Quantum cognitive or learning models require sustained phase coherence; however, interaction with the environment induces rapid decoherence, destroying superposition and entanglement and thereby limiting computational depth and reliability. The warm, wet, noisy environment of biological brains seems particularly hostile to quantum coherence. Fault-tolerant computation demands substantial redundancy through error-correcting codes, dramatically increasing physical qubit requirements and imposing nontrivial architectural and control complexity. Surface codes, for example, require large physical-to-logical qubit ratios. Current hardware platforms face significant barriers in scaling to the millions of high-fidelity qubits likely required for practically meaningful quantum learning or AGI-relevant workloads. No existing platform demonstrates a clear path to this scale with required coherence times. Realistic devices operate in the noisy intermediate-scale quantum (NISQ) regime, where gate errors, readout noise, and limited circuit depth constrain achievable algorithmic advantage and reproducibility. NISQ-era algorithms may not outperform classical approaches for cognitively relevant tasks. A unified theoretical framework characterizing sample complexity, generalization bounds, and provable speedups for quantum machine learning remains underdeveloped.

Many claimed advantages are contingent on strong oracle assumptions that may not hold in practice. There is no mathematically rigorous, consensus definition of Artificial General Intelligence, complicating claims that quantum methods meaningfully advance toward AGI rather than providing task-specific acceleration. Without clear metrics, progress assessment remains subjective. Given foundational and engineering limitations, strong claims about quantum-enhanced cognition or AGI must be framed as speculative. Intellectual rigor requires distinguishing formal theoretical possibility from experimentally validated capability.

So, the question is: what can be a feasible research roadmap for this? In the near term (0-5 years), we can look at developing integrated systems where neuromorphic hardware performs energy-efficient spiking and attractor-based computation, while quantum co-processors handle combinatorial search, sampling, or high-dimensional state evolution, with well-defined quantum-classical interfaces. We can systematically extract algorithmic principles from quantum formalisms—tensor networks, amplitude-based similarity metrics, quantum walk dynamics—to design classically tractable approximations that retain structural advantages without requiring fault-tolerant hardware. We can also implement reduced-scale cognitive modules—associative memory, reinforcement learning primitives, attention-like selection mechanisms—on noisy intermediate-scale quantum platforms to empirically evaluate feasibility and resource scaling.

When it comes to the medium-term (5-10 years), we can formulate Bayesian networks and probabilistic inference mechanisms within Hilbert-space representations, leveraging quantum amplitude encoding for compact joint distribution representation and potentially accelerated sampling or marginalization. We can recast predictive coding and active inference in terms of state vectors and operators in complex vector spaces, enabling unified treatment of belief updating, uncertainty, and hierarchical generative modeling under linear-algebraic dynamics. We can also establish rigorous complexity-theoretic benchmarks, noise models, and comparative baselines to distinguish genuine quantum advantage from classical stimulability across cognitive and learning tasks. Finally, for the long-term, beyond 10 years, we can position this agenda as a staged program progressing from quantum-inspired classical methods and small-scale NISQ demonstrations toward fault-tolerant hybrid architectures, contingent on advances in hardware scalability, error correction, and theoretical learning guarantees. The ultimate goal is synthetic intelligence beyond classical limits, where intelligence is analyzed through representational capacity and state-space geometry rather than solely through scaling parameters and datasets.

Quantum Approaches to Consciousness

This brings us to what philosopher David Chalmers famously called the "hard problem of consciousness": How do electrochemical signals in our brains transform into our rich inner mental life? While we can trace neural pathways and map brain activity, explaining how these physical mechanisms create subjective experiences—the redness of red or the taste of chocolate—remains remarkably challenging. Quantum mechanics, with its strange and counterintuitive properties, has emerged as an intriguing lens through which to examine consciousness. Some scientists suggest that quantum phenomena occurring in brain cells might help bridge the gap between physical processes and conscious experience. While these ideas remain speculative, they highlight how understanding consciousness may require thinking beyond classical physics.

Quantum mind theories represent an ambitious attempt to bridge neuroscience and quantum physics, suggesting that quantum mechanical processes in the brain might be key to understanding consciousness. These hypotheses propose that quantum phenomena—superposition, entanglement, and wave-function collapse—could explain how the brain generates conscious experience. However, these theories currently exist in a scientific gray area. While proposed by some respected physicists and neuroscientists, they remain largely theoretical and unproven. The main challenge lies in demonstrating how quantum effects, typically observable only at the microscopic level, could meaningfully influence the warm, wet environment of the brain. There is an important distinction between legitimate quantum mind theories, which attempt to apply rigorous physics to consciousness, and quantum mysticism—pseudoscientific ideas that misuse quantum terminology for unfounded spiritual or metaphysical claims. The key is focusing on testable hypotheses based on established quantum mechanics rather than speculative interpretations.

Quantum approaches to consciousness can be categorized into three main types. The first kind is where *consciousness is seen as a manifestation of quantum processes in the brain:* theories proposing that specific quantum mechanisms in neural structures give rise to conscious experience. The second kind is where we *use quantum concepts to understand conscious mental activity without referring to the brain:* approaches applying quantum mathematical formalism to model mental phenomena, irrespective of underlying neural implementation. The third and final kind is where *matter and consciousness are seen as dual aspects of one underlying reality:* dual-aspect monist frameworks where mental and physical are two manifestations of a psychophysically neutral ground. Quantum biology provides the most empirically grounded foundation for quantum approaches to consciousness, with certain proposed examples of quantum effects in biological systems: *photosynthesis* that uses quantum coherence to achieve remarkable energy transfer efficiency, *bird navigation* that may involve quantum entanglement in magnetoreceptors, and *enzyme catalysis* that potentially utilizes quantum tunneling for certain reactions. These demonstrated quantum biological phenomena establish a precedent for macroscopic biological effects arising from quantum mechanisms, though their relevance to neural computation remains speculative.

The Orchestrated Objective Reduction (Orch-OR) theory, developed by Roger Penrose and Stuart Hameroff, proposes that consciousness arises from quantum computations in brain neuron microtubules. The primary key element is *Objective Reduction (OR):* A process where quantum coherence grows in microtubules until reaching a critical threshold related to quantum gravity, at which point it self-collapses. Penrose proposes that the critical threshold arises from instability in Planck-scale separations (superpositions) in spacetime geometry, linking collapse to fundamental physics. Another key element is the concept of *orchestration:* Connective proteins, such as microtubule-associated proteins (MAPs), influence qubit state reduction by modifying the spacetime separation of superimposed states, "orchestrating" the collapse process. The theory suggests that output states following objective reduction are influenced by a non-computable factor ingrained in fundamental spacetime, linking to Penrose's interpretation of Gödel's theorem. Consciousness (Orch-OR) is proposed to occur during self-collapse when quantum computation in coherent superposition undergoes objective reduction. Microtubule-associated proteins "tune" quantum oscillations and influence collapse outcomes. The theory proposes quantum entanglement and synchronization of neurons via microtubules, with tubulin protein inside microtubules containing pi electrons whose disruption by anaesthetics may explain the loss of consciousness. To address the challenge of the brain being warm, wet, and noisy, the theory suggests that microtubules can isolate and protect quantum processes. While generating testable predictions about microtubule vibrations and consciousness, OrchOR remains highly controversial, with debates about anaesthetic-tubulin interactions and experimental evidence.

Ricciardi and Umezawa proposed using Quantum Field Theory (QFT) to describe brain states, particularly memory, conceiving memory states as vacuum states of quantum fields. QFT, unlike standard quantum mechanics, handles systems with infinitely many degrees of freedom. Spontaneous symmetry breaking in QFT generates Nambu-Goldstone bosons, producing long-range correlations and ordered patterns in neuronal assemblies. Umezawa proposed that coherent neuronal assemblies linked to memory are vacuum states, activated by external stimuli for conscious recollection. Vitiello incorporated dissipation through system-environment interaction, leading to multiple coded vacuum states and temporally limited memory. Dissipation generates a time arrow and entanglement within the system. Freeman and Vitiello clarified that the QFT model describes the brain's physical states, not mental states themselves, studying neurobiological observables and finding evidence for non-equilibrium phase transitions. Helias and colleagues used methods from quantum field theory to confirm critical points in brain dynamics models. Yearsley and Pothos proposed using quantum probability theory to assign probabilities to cognitive variable precision over time.

Quantum Brain Dynamics applies quantum field theory to brain function, suggesting that long-range coherent waves (like Bose-Einstein condensates) facilitate neural activity and memory processes. Memory is proposed to be stored non-locally as a result of coherent quantum states, allowing dynamic retrieval and association beyond classical neural network models. QBD posits a two-level model where the classical brain (neural firings) interacts with a deeper quantum substrate (coherent wave dynamics), enabling complex cognitive functions. The model incorporates dissipative dynamics (Vitiello's approach), where brain-environment interactions create a thermal bath stabilizing memory and consciousness.

As Nishiyama et al. (2021) articulate: "Higher cognitive functions including memory, perception, attention, decision-making, learning and consciousness are associated with synchronized oscillations of large neuronal groups, known as neuronal synchrony... In terms of physical properties, memory can be characterized as having the property of diversity, long-term but imperfect stability, and nonlocality. Memory storage appears distributed and holographic, encoded at various scales and locations. Scale-free architectures, including fractal-like "grid cells" mapping spatial location (Nobel Prize 2014 to Moser, Moser, and O'Keefe), exhibit self-similar patterns repeating at multiple spatio-temporal levels. This scale-free property means memories do not disappear when particular neuron domains are damaged—a nonlocal nature that conventional neuroscience struggles to explain.

Researchers have used modified MRI to detect potential entanglement between proton spins in the brain and heart activity. Correlating brain activity with heartbeat-induced potential (HEP), observed signals suggest a non-classical intermediary might mediate this entanglement. The presence of this signal only when subjects were conscious indicates that consciousness may emerge as a feature of quantum entanglement, suggesting brain functions must operate non-classically. We then have Stapp's approach, which builds on von Neumann's quantum mechanics, particularly measurement and quantum state reduction. He locates the interface between observed and observing systems within the observer's brain, inspired by von Neumann's freedom to choose this boundary. Extending beyond the Copenhagen interpretation, Stapp considers Heisenberg's "potential" as reality independent of measurement. Each actual event has an experiential aspect—a "feel"—contributing to intrinsic actuality, introducing subjective elements. Drawing inspiration from Whitehead, Stapp views mind-matter relationships as psychophysically neutral potential antecedents. Conscious intentions can influence brain activity not through direct interaction but via constraints imposed by mind-matter correlations. Each conscious experience has a physical counterpart: quantum state reduction creating specific brain activity patterns—the neural correlate. Intentional mental effort can prolong neuronal assembly activity due to quantum Zeno-type effects.

Beck and Eccles proposed one of the first concrete models of quantum mechanics in brain processes, focusing on information transfer at synaptic clefts—areas affected by anaesthetics that switch off consciousness. Their model centers on exocytosis (neurotransmitter release) in synapses, showing quantum processes dominate over thermal processes at energies above 10^{-2} eV and masses below 10 electron masses, operating at ultrafast timescales of 10^{-12} seconds. The model proposes a quantum trigger mechanism using quasi-particles and tunneling processes, successfully predicting exocytosis probability ranges between 0 and 0.7, matching empirical observations. Eccles suggested this quantum mechanism could allow mental causation—consciousness influencing brain states by modulating exocytosis probability—though this remains speculative. A key challenge is explaining how random quantum processes at individual synapses could generate coherent neural assemblies associated with conscious thought, though Beck suggested quantum stochastic resonance as a possible solution. A more recent proposal by Fisher focuses on Posner molecules (calcium phosphate), suggesting their phosphate nuclear spins could serve as protected quantum bits, potentially maintaining coherence for hours or days and influencing neurotransmitter release across multiple neurons. Fisher's theory is testable—replacing calcium with different lithium isotopes should affect quantum decoherence times and neural activity, with some supporting evidence from animal studies and connections to lithium's therapeutic effects in bipolar disorder.

The quantum behavior of mental systems can be studied through a state space description without involving quantum brain processes. Mental states are defined by partitions of neural state spaces—partitions carefully constructed to create robust mental state definitions. A key insight is that quantum-like features in mental systems don't necessarily require quantum brain processes. These features can emerge from how we partition and describe classical neural states into mental states. Non-commuting operations and non-Boolean logic naturally appear in mental systems because: (1) the order of mental operations affects outcomes, and (2) mental propositions often have varying degrees of truth rather than simple yes/no values. Pylkkänen proposed using intuitive understanding of these mental quantum features to better grasp quantum physics itself, reversing the usual direction of explanation from physics to mind. This framework for generalizing quantum theory beyond physics has proven valuable because it maintains clear connections to conventional quantum physics while producing detailed theoretical and empirical results in psychological applications. Seven main areas have been studied using this quantum mental features approach. The first is decision processes. Quantum probability models explain decision outcomes, conjunction/disjunction effects, and confidence dynamics better than classical models. They clarify long-standing cognitive biases and demonstrate quantum interference effects in decision-making, providing mathematical frameworks for phenomena that resist classical explanation. We then have order effects. Quantum models explain why question order influences survey responses, revealing non-classical contextual dependencies. This provides a fundamental covariance condition for such effects, capturing how measurement context shapes response distributions.

We then have the concept of bistable perception. The Necker-Zeno model links bistable perception to the quantum Zeno effect, predicting psychophysical time scales. It also suggests violations of temporal Bell inequalities, implying temporally nonlocal cognitive states—a striking prediction for empirical investigation. Small recurrent networks can exhibit non-commutative learning, where input recognition depends on presentation order. This quantum-like structure of learning has implications for educational practice and cognitive modeling. Quantum-inspired models explain context-dependent meaning formation and entanglement-like features in concept combinations. This challenges the classical view of separable word meanings and supports holistic semantic representations. Briegel and Müller (2015) propose a quantum model where agents develop autonomy via non-deterministic quantum random walks. This extends quantum reinforcement learning (Paparo et al., 2012) to explain emergent intentionality, suggesting quantum formalism may capture essential features of agency. Beyond standard quantum entanglement, super-quantum correlations challenge classical and quantum limits. Methods help distinguish genuine quantum correlations from classical contextual effects, providing tools for rigorous empirical investigation.

Mind and Matter as Dual Aspects of Underlying Reality

Dual-aspect theories propose that the mental and physical are two aspects of a single, underlying reality that is neither mental nor physical. These theories contrast with mind-body dualism, which posits fundamentally distinct substances for mind and matter. In dual-aspect monism, the distinction between mind and matter arises from an epistemic split—a way of gaining knowledge about the underlying reality. One class of dual-aspect models, exemplified by Russell and Chalmers, sees the mental and physical as reducible to compositional arrangements of neutral elements. Another class, including models by Pauli, Jung, Eddington, Wheeler, Bohm, and Hiley, views the underlying domain as holistic, with mental and physical aspects emerging through symmetry breaking. Eddington sought a basis reality, the "spiritual world," free from the mind-matter distinction, linking physical objects and phenomenal perception. Wheeler's "it from bit" suggests that question selection partitions reality, enabling bit definition and introducing meaning circuits connecting objective registration and subjective selection. Bohm and Hiley use implicate and explicate order concepts, with mental and physical states unfolding from psychophysically neutral implicate order termed "holomovement." Dual-aspect thinking invites interpretation in the spirit of panpsychism—the doctrine that mind is a fundamental world feature existing throughout the universe. However, these theories are based on speculations exceeding the scope of contemporary quantum theory.

The Pauli-Jung Conjecture proposes a dual-aspect monist view where mind and matter are aspects of a psychophysically neutral (PPN) reality. This conjecture explores correlations between mental and physical without direct causal interaction, respecting causal closure. It draws on quantum theory concepts like complementarity, non-locality, and the observer effect. The PPN reality contains archetypes that act as "ordering factors," influencing mental and physical manifestations. These archetypes are not composed of elements but conceived holistically. Inspired by quantum entanglement, it posits that quantum correlations may model the mind-matter relationship. The mental side involves the personal unconscious, while the physical side involves entangled quantum states.

The Pauli-Jung conjecture explores synchronicity—meaningful, acausal connections between internal psychological events and external world events. Synchronistic events are considered particularly striking examples of mind-matter correlations within dual-aspect monism. Meaning is an essential criterion for synchronistic correlations, highlighting its significance in understanding mind-matter relationships. Jung viewed synchronicity as an acausal connecting principle complementing causality in understanding the psyche and world. Jung and Pauli drew on quantum theory elements, such as entanglement and non-locality, to explore mind-matter relationships. Atmanspacher and Fach proposed a comprehensive taxonomy of mind-matter correlations, classifying "exceptional experiences" based on deviations from conventional reality models. Contemporary physicists have suggested quantum entanglement may model mind-matter relationships within the Pauli-Jung framework. The conjecture posits that archetypes, residing in the collective unconscious, act as ordering factors influencing both mental and physical manifestations. A 2016 study indicated 70% of therapists found synchronicity experiences potentially useful in therapy.

Physics seeks to unify fundamental forces and particles, hinting at simpler underlying principles. As physics unifies different forces, it shows their interconnectedness in the world. Unification may lead to descriptions with fewer fundamental entities and interactions, suggesting more parsimonious reality at deeper levels. The discovery of relationships between previously disparate phenomena, like electricity and magnetism, demonstrates intrinsic correlative tendencies in nature. As physics peels back layers of complexity, it could potentially reveal a level of reality approaching the concept of *Cit* in Indian philosophy—primordial, pure consciousness that is the ground of all being, a unified field of awareness from which everything arises. While speculative, unification's drive toward simplicity, combined with observed correlations in nature, offers a speculative bridge to the idea of fundamental consciousness underlying all existence.

Conclusion

The central hypothesis emerging from this investigation is that progress toward general intelligence may depend less on increasing data volume and more on enriching the mathematical structure in which hypotheses, uncertainties, and counterfactuals are encoded. Biological cognition appears to exploit structured uncertainty via probabilistic inference, hierarchical generative models, and efficient coding, leveraging environmental regularities rather than brute-force data accumulation. Contemporary AI systems primarily achieve performance gains through large-scale optimization over high-dimensional parameter spaces, relying on massive datasets to approximate complex input-output mappings. Quantum computation operates over exponentially large Hilbert spaces, where superposition, interference, and entanglement enable structured manipulation of possibility distributions rather than explicit enumeration of alternatives. The strategic implication is clear: expanding the geometry and algebra of representational spaces—potentially via quantum or quantum-inspired formalisms—may unlock qualitatively different regimes of inference, search, and generalization. As stated in the closing principle: *"General intelligence may not require more data—it may require a richer representation of possibility."* This vision positions quantum approaches to AGI not as a replacement for classical methods but as a complementary paradigm addressing fundamental limitations in representational capacity and computational efficiency.

The path forward requires interdisciplinary collaboration spanning quantum physics, computer science, neuroscience, and cognitive psychology, grounded in rigorous experimental validation and theoretical clarity about what quantum methods can and cannot offer. The journey from today's narrow AI to tomorrow's general intelligence will likely require multiple paradigm shifts, of which quantum computing is one promising candidate. Whether quantum principles ultimately prove essential for AGI or simply inspire more powerful classical algorithms, the exploration itself advances our understanding of both computation and cognition.

References

Atmanspacher, H., and Filk, T. (2006). Complexity and non-commutativity of learning operations on small recurrent networks. *BioSystems*, 85(1), 84-93.

Atmanspacher, H., and Filk, T. (2013). The Necker-Zeno model for bistable perception. *Topics in Cognitive Science*, 5(4), 800-817.

Atmanspacher, H., and Römer, H. (2012). Order effects in sequential measurements of noncommuting psychological observables. *Journal of Mathematical Psychology*, 56(4), 274-280.

Beck, F., and Eccles, J. C. (1992). Quantum aspects of brain activity and the role of consciousness. *Proceedings of the National Academy of Sciences*, 89(23), 11357-11361.

Beer, K., Bondarenko, D., Farrelly, T., et al. (2020). Training deep quantum neural networks. *Nature Communications*, 11(1), 808.

Bell, J. S. (1964). On the Einstein-Podolsky-Rosen paradox. *Physics Physique Fizika*, 1(3), 195-200.

Bennett, C. H., Brassard, G., Crépeau, C., et al. (1993). Teleporting an unknown quantum state via dual classical and Einstein-Podolsky-Rosen channels. *Physical Review Letters*, 70(13), 1895-1899.

Bennett, C. H., and Wiesner, S. J. (1992). Communication via one- and two-particle operators on Einstein-Podolsky-Rosen states. *Physical Review Letters*, 69(20), 2881-2884.

Briegel, H. J., and Müller, T. (2015). A chance for attributable agency. *Minds and Machines*, 25(3), 261-279.

Bruza, P. D., Kitto, K., Ramm, B. R., and Sitbon, L. (2015). A probabilistic framework for analysing the compositionality of conceptual combinations. *Journal of Mathematical Psychology*, 67, 26-38.

Busemeyer, J. R., Wang, Z., and Townsend, J. T. (2006). Quantum dynamics of human decisionmaking. *Journal of Mathematical Psychology*, 50(3), 220-241.

Dang, Y., Jiang, N., Hu, H., Ji, Z., and Zhang, W. (2018). Quantum k-nearest neighbor algorithm. *Quantum Information Processing*, 17(4), 1-18.

Durr, C., and Hoyer, P. (1996). A quantum algorithm for finding the minimum. *arXiv preprint quant-ph/9607014*.

Dzhafarov, E. N., and Kujala, J. V. (2013). Quantum entanglement and the issue of selective influences in psychology: An overview. *PLoS One*, 8(5), e61712.

Ekert, A. K. (1991). Quantum cryptography based on Bell's theorem. *Physical Review Letters*, 67(6), 661-663.

Fisher, M. P. (2015). Quantum cognition: The possibility of processing with nuclear spins in the brain. *Annals of Physics*, 362, 593-602.

Gabora, L., and Aerts, D. (2002). Contextualizing concepts using a mathematical generalization of the quantum formalism. *Journal of Experimental and Theoretical Artificial Intelligence*, 14(4), 327-358.

Nishiyama, A., Tanaka, S., and Tuszynski, J. A. (2021). Nonlocality and stability of memory storage in quantum brain dynamics. *Physica A: Statistical Mechanics and its Applications*, 567, 125706.

Paparo, G. D., Dunjko, V., Makmal, A., et al. (2012). Quantum speedup for active learning agents. *Physical Review X*, 4(3), 031002.

Popescu, S., and Rohrlich, D. (1994). Quantum nonlocality as an axiom. *Foundations of Physics*, 24(3), 379-385.

Srikumar, M., Hill, C. D., and Hollenberg, L. C. (2024). A kernel-based quantum random forest for improved classification. *Quantum Machine Intelligence*, 6(1), 10.

Suzuki, T., Hasebe, T., and Miyazaki, T. (2024). Quantum support vector machines for classification and regression on a trapped-ion quantum computer. *Quantum Machine Intelligence*, 6(1), 31.

Wang, Z., Solloway, T., Shiffrin, R. M., and Busemeyer, J. R. (2014). Context effects produced by question orders reveal quantum nature of human judgments. *Proceedings of the National Academy of Sciences*, 111(26), 9431-9436.

Systems Engineering for Intelligence: A Transdisciplinary Educational Approach

Dr. Vishwanathan Ramesh

Goethe University, Center for Cognition and Computation, Frankfurt, Germany

Abstract

As artificial intelligence transforms society, educational institutions face the challenge of preparing the next generation of computer scientists and engineers. This article presents Goethe University Frankfurt's innovative systems engineering curriculum that bridges traditional engineering approaches with cutting-edge artificial intelligence. At its core is the "(C, T, P) -+ (HW/SW Sys)" framework—a systematic mapping from Context, Tasks, and Performance requirements to Hardware/Software System designs. Drawing from a 14-year journey of research and educational development, this transdisciplinary approach combines systems thinking with state-of-the-art machine learning to train "holistic systems thinkers" who can develop AI systems that are safe, explainable, and effective. The curriculum's evolution through the AISEL (AI Systems Engineering Laboratory) and ALI (AI and Digital Technologies in Learning and Instruction) initiatives demonstrates how systems engineering principles can address complex challenges at the intersection of technology, science, and society. This article presents the vision, pedagogical approach, and demonstrated impact of this educational framework, which responds directly to Germany's critical shortage of MINT professionals while preparing students for the unique challenges of the AI revolution era.

Introduction: The AI Revolution and Educational Response

As we navigate through what many call the "AI revolution," society faces unprecedented technological transformation. Systems that once seemed firmly in the realm of science fiction—machines that can converse naturally, create art, write code, and reason about complex problems—are now part of our everyday reality. This rapid evolution raises profound questions: How should we design these increasingly capable systems? What skills will the next generation need to navigate this landscape? And perhaps most importantly, how can we ensure that AI evolves safely alongside humanity? At Goethe University Frankfurt, these questions have driven the development of a unique educational approach that bridges traditional systems engineering with cutting-edge artificial intelligence. This approach, centered on what Prof. Dr. Visvanathan Ramesh calls "Systems Engineering for Intelligence," represents a comprehensive response to the educational challenges of the AI era.

The Three Waves of AI

To understand the context of our educational framework, we must first recognize the historical evolution of artificial intelligence, which can be conceptualized as occurring in three distinct waves: The first wave is called handcrafted knowledge. Systems based on explicitly programmed rules and knowledge bases, characterized by high transparency but limited flexibility. The second wave is statistical Learning. Data-driven approaches, including deep learning, offering impressive performance but often functioning as black boxes with limited explainability. The third wave is contextual adaptation. The emerging frontier focuses on systems that combine the strengths of both approaches—performance and adaptability with transparency and explainability. Our educational approach is particularly focused on preparing students for this third wave—training engineers who can design AI systems that are not only technically sophisticated but also transparent, explainable, and aligned with human values.

Defining Intelligence: "Productive-Laziness"

At the heart of our approach is a pragmatic definition of intelligence that guides our educational philosophy. Drawing from philosopher Aaron Sloman's insight, we define intelligence as "Productive-Laziness" or "Lazy Productiveness"—the ability to solve complex problems efficiently by understanding which aspects require detailed attention and which can be simplified or abstracted away. This definition has profound implications for AI system design. It suggests that truly intelligent systems must balance powerful computational capabilities with efficient resource allocation, much as human intelligence combines intuitive, fast thinking with deliberative, slow thinking. Our educational approach seeks to develop this balanced perspective in students, training them to design systems that embody this principle of productive laziness.

The MINT Gap and Societal Needs

Our educational framework responds directly to a critical societal challenge: Germany's shortage of over 300,000 MINT (Mathematics, Informatics, Natural Sciences, Technology) professionals. This shortage affects not just the technology sector but all aspects of an increasingly digitalized economy. By developing a transdisciplinary approach that bridges traditional disciplinary boundaries, we aim to train a new generation of "holistic systems thinkers" who can address complex socio-technical challenges. These professionals combine technical depth with systemic breadth—understanding both the mathematical foundations of AI and its broader implications for society.

The Genesis of a Vision: From Industry to Academia

The systems engineering curriculum at Goethe University represents the culmination of a 14-year journey that bridges industry experience with academic research. To understand the distinctive nature of this approach, we must trace its development from its industrial origins to its current academic implementation.

Industrial Foundations (1995-2011)

The foundations of our approach were developed during Prof. Ramesh's tenure at Siemens Corporate Research, where he led the development of real-time vision systems for various domains including security, mobility, healthcare, and energy. This work required addressing fundamental challenges in systems design:

- How to systematically map from application contexts and requirements to system configurations
- How to balance model-based design with data-driven machine learning
- How to ensure robustness and safety in complex, real-world deployments
- How to organize large, distributed teams around a coherent systems vision

This industrial experience yielded several key insights that would later inform the educational approach:

- System configuration is a function of context, tasks, and requirements
- Effective system architectures often mirror cognitive processes, combining fast "reflexive" components with more deliberative "reflective" components
- Hybrid designs that combine model-based approaches with machine learning often outperform purely data-driven or purely model-based approaches

These insights were encapsulated in what became known as the "Common Visual Cognition Framework," a systems architecture that addressed predictability, scalability, modularity, and extensibility in vision systems.

Transition to Academia (2011-Present)

In 2011, Prof. Ramesh transitioned to Goethe University Frankfurt, bringing this industry-tested systems perspective into academic research and education. This transition coincided with the explosive growth of deep learning and the second wave of AI, creating both opportunities and challenges for systems engineering education. The academic journey at Goethe University has unfolded through several key phases:

- Bernstein Focus in Neuro Technology (2011-2016) - Building transdisciplinary foundations connecting brain science with systems engineering
- AISEL: AI Systems Engineering Laboratory (2019-2021) - Formalizing holistic systems training with a four-perspective framework
- ALI: AI & Digital Technology in Learning (2022-2025) - Adapting systems thinking to educational technology and integrating generative AI

Throughout this evolution, the central organizing principle has remained consistent: training "holistic systems thinkers" who can translate contexts, tasks, and performance requirements into effective AI system designs.

The Core Framework: (C, T, P) ---- (HW/SW Sys)

At the heart of our educational approach lies a unifying framework that serves as the central organizing principle across all courses and projects. This framework represents the systematic mapping from Context, Tasks, and Performance requirements (C, T, P) to Hardware/Software Systems designs (HW/SW Sys).

Understanding the Framework Components

Context (C)

The environment in which the system operates, including:

- Physical constraints and properties
- User needs and characteristics
- Regulatory and ethical considerations
- Domain-specific knowledge and norms

Tasks (T)

The specific functions the system must perform, including:

- Primary objectives (e.g., classification, prediction, control)
- Secondary objectives (e.g., monitoring, reporting, adaptation)
- Task decomposition and prioritization
- Success criteria for each task

Performance Requirements (P)

Metrics that define successful operation, including:

- Accuracy and precision targets
- Speed and latency constraints
- Resource utilization limits
- Safety and reliability thresholds
- Explainability and transparency requirements

Hardware/Software Systems (HW/SW Sys)

The implementation architecture, including:

- Hardware components and configurations
- Software architectures and algorithms
- Integration patterns and interfaces
- Deployment and operation strategies

This mapping represents the fundamental challenge of systems engineering: translating real-world needs into technical solutions through a systematic process of modeling, analysis, implementation, and validation.

Four Perspectives on Systems Design

To operationalize this framework, our curriculum teaches students to view systems from four complementary perspectives:

1) **User Perspective** - Understanding the application context, stakeholder needs, and acceptance criteria
2) **Modeling Perspective** - Developing formal representations of the problem domain using scientific and engineering models
3) **Implementation Perspective** - Translating models into executable software and hardware configurations
4) **Validation Perspective** - Evaluating system performance against requirements through testing, analysis, and user feedback

By systematically integrating these perspectives, students learn to address the full lifecycle of AI systems development, from initial requirements to deployment and maintenance.

Systems Design as Cognitive Architecture

A distinctive feature of our approach is the integration of cognitive science principles into systems design. Drawing from dual-system models in psychology, we teach students to design systems that combine:

- Fast, reflexive components that provide real-time responses based on learned patterns
- Slower, reflective components that perform more deliberative reasoning and detailed estimation
- Feedback mechanisms that update world knowledge based on system performance

- Control mechanisms that adjust sensing and processing parameters based on context

This cognitive architecture approach helps students design systems that can efficiently allocate computational resources—embodying the principle of "productive laziness" that defines intelligence in our framework.

Integrated Course Structure

Our curriculum consists of complementary courses that approach AI systems engineering from different perspectives, providing a comprehensive educational experience that balances theoretical understanding with practical implementation skills.

Core Courses and Their Integration

Systems and Software Engineering I & II

These courses focus on software architecture, design patterns, and implementation processes for large-scale software systems. Key topics include:

- Software lifecycles and development methodologies
- Requirements engineering and specification
- Architecture and design patterns
- Quality assurance and testing strategies
- Project management and team dynamics

The courses emphasize implementation and execution aspects of building large software systems, with special attention to emerging paradigms like "Software 2.0" that incorporate machine learning components.

Systems Engineering Meets Life Sciences I & II

These courses explore how intelligent systems (with special emphasis on vision systems) can be formally specified, designed, and implemented. They highlight connections to biologically inspired architectures and cover:

- Formal specification of context, tasks, and performance requirements

- Biologically inspired computational architectures
- Integration of model-based and data-driven approaches
- System validation and performance characterization

These courses provide the conceptual foundation for the "(C, T, P) ---t (HW/SW Sys)" mapping framework.

Machine Learning I & II

These courses cover both theoretical foundations and practical implementation of machine learning algorithms, with emphasis on the interplay between model-based and data-driven approaches. Key topics include:

- Fundamentals of supervised and unsupervised learning
- Deep learning architectures and training methodologies
- Probabilistic graphical models and Bayesian inference
- Geometric deep learning and specialized architectures
- Transformers and large language models
- Uncertainty quantification and formal verification

Machine Learning II specifically addresses advanced topics, including the five tribes of machine learning (Symbolists, Connectionists, Evolutionaries, Bayesians, and Analogizers), bridging philosophical foundations with practical implementations.

Practice Course in Machine Learning

This course provides hands-on experience with real-world projects and applications, allowing students to apply theoretical knowledge to practical challenges. Projects typically involve:

- Dataset preparation and exploration
- Model selection and implementation
- Hyperparameter optimization

- Performance evaluation and comparison
- Deployment considerations

Seminar on Pattern Analysis and Machine Intelligence

This seminar integrates concepts from machine learning, cognitive science, and philosophy of AI, encouraging students to engage deeply with foundational questions about intelligent systems. Topics often include:

- Historical and philosophical perspectives on AI
- Connections between biological and artificial intelligence
- Ethical considerations in AI system design
- Current research frontiers and debates

Pedagogical Integration

These courses are not taught in isolation but are designed to reinforce and complement each other. Integration is achieved through:

- Common conceptual framework (the (C, T, P) ---t (HW/SW Sys) mapping)
- Shared case studies and application examples
- Cross-referenced assignments and projects
- Unified terminology and notation
- Team-teaching across disciplines

This integrated approach ensures that students develop a coherent understanding of AI systems engineering that transcends traditional disciplinary boundaries.

Project-Based Learning: AISEL and ALI

Project-based learning forms the core of our educational approach, allowing students to apply theoretical concepts to real-world challenges. This approach is exemplified by two major initiatives: AISEL (Artificial Intelligence Systems Engineering Laboratory) and ALI (AI and Digital Technologies in Learning and Instruction).

AISEL: AI Systems Engineering Laboratory

The AISEL project (2019-2021) trained computer scientists and engineers as "holistic systems thinkers" capable of translating contexts, tasks, and performance requirements into AI system designs. Key components included:

- Student Projects
- AISEL supported a range of student projects that applied the (C, T, P) ---t (HW/SW Sys) framework to real-world challenges:
    - Smart Classroom - Using air quality sensors to optimize learning conditions through a system that monitors CO2 levels and alerts when ventilation is needed
    - AI Sudoku - Developing a system to detect and solve Sudoku puzzles in newspaper images, including robustness to paper crumpling
    - Greenhouse Yield Prediction - Creating systems to monitor plant growth and predict yields, including segmentation of leaves and measurement of photosynthesis areas
    - Analysis of Learning Dynamics in Deep Networks - Investigating how neural networks learn and which data instances are learned first, regardless of architecture
    - Simulation for Performance Assessment of Continual Learning - Developing a procedural world generation framework for systematic evaluation of continuous machine learning systems

Industrial Collaborations

AISEL established partnerships with companies including Daimler AG, Volkswagen AG, P3 Group, and AIGrow. These collaborations provided:

- Real-world use cases for student projects
- Industry mentorship and feedback
- Validation of curriculum relevance
- Career pathways for graduates

Key Competencies Development

AISEL focused on developing a range of competencies essential for AI systems engineers:

- Soft skills for leadership and communication
- Bias avoidance in AI system design
- Transparency and explainability
- Robustness and resilience
- Architecture optimization
- Adaptive behavior implementation
- Privacy protection
- Safety certification

These competencies were developed through a combination of coursework, project experiences, and interactions with industry partners.

ALI: AI and Digital Technologies in Learning and Instruction

Building on AISEL's foundation, the ALI project (2021-2025) applies systems engineering principles specifically to educational technology:

Interdisciplinary Master's Program

ALI developed a master's program that integrates psychology, computer science, and educational technologies. The program follows a three-phase structure:

1) First Phase (first semester) - Courses on psychology, computer science fundamentals, and theoretical foundations of human and machine intelligence
2) Second Phase (second and third semesters) - Focus on socio-technical systems for education, covering design, evaluation, and implementation
3) Final Phase (internship and thesis) - Application of knowledge in practical educational settings and research

GenAI and LLMs in Education

A key focus of ALI is understanding the implications of generative AI and large language models for educational technology:

- Exploring LLMs as creative tools for educators and students
- Investigating LLMs as sequence-to-sequence transformation architectures
- Developing and evaluating LLM-based applications in education
- Conducting experiments on LLM use in computer science education

The project has explored specific applications including programming assistance, flipped classroom materials creation, exam question preparation, and content summarization.

Affective AI in Educational Scenarios

ALI also investigates affective AI—technologies that understand and respond to students' emotional needs:

- Understanding the role of emotions and engagement in learning
- Developing systems that can interpret emotional states in educational contexts
- Addressing challenges in evaluating affective AI, given the subjective nature of emotion
- Creating frameworks for integrating affective AI into educational platforms

Systems Engineering Perspective

ALI applies the same four-perspective framework developed in AISEL to educational technology:

- User view - Understanding educator and learner needs
- Modeling view - Developing contextual models of learning
- Implementation view - Creating component libraries and frameworks

- Evaluation view - Validating systems against learning outcomes

This consistent framework demonstrates how systems engineering principles can be applied across domains, from computer vision to educational technology.

Impact and Outcomes

The systems engineering curriculum at Goethe University has demonstrated significant impact across academic, industrial, and societal dimensions.

Academic Impact

Educational Outcomes

The curriculum has produced impressive educational results:

- 4 PhD dissertations completed at Goethe University
- 33 Master's and Bachelor's theses completed
- Alumni pursuing academic careers, including Martin Mundt becoming a W3 Professor for Continual Learning at the University of Bremen

Research Contributions

The curriculum has supported research advances in several areas:

- Simulation frameworks for cognitive vision
- Design methodologies for continual learning in deep neural networks
- Cognitive vision architectures for real-time applications
- Approaches to transparent deep learning models
- Procedural world generation for systematic evaluation of machine learning

Industrial Impact

Industry Placements

- Graduates of the program have assumed significant roles in industry:
- Michael Rammensee (Postdoc in AISEL) became CEO of AI Quality and Test Hub GmbH
- Subbu Veerasavarappu is at Verisk Analytics GmbH
- Other students placed at Continental, Bosch Research, Accenture, Deutsche Bahn, and regional startups

Industry Collaborations

The program has established productive partnerships with leading companies:

- Joint projects with Daimler AG on automotive self-localization using polarimetric radar
- Collaboration with Volkswagen AG on traffic sign detection with adversarial perturbations
- Partnership with AIGrow on greenhouse monitoring and yield prediction
- Strategic support to the state of Hessen and VDE in establishing AI Quality and Test Hub GmbH

Societal Impact

Addressing the MINT Gap

The curriculum directly addresses Germany's critical shortage of MINT professionals by:

- Training interdisciplinary professionals who bridge traditional boundaries
- Developing both technical depth and systemic breadth in graduates
- Creating pathways for continuous learning and adaptation

Advancing Responsible AI

The curriculum promotes responsible AI development through:

- Emphasis on transparency and explainability in AI systems
- Integration of ethical considerations throughout the design process
- Focus on safety certification and validation methodologies
- Training in bias avoidance and fairness principles

From Systems Engineering to Transdisciplinary Computing Education

The systems engineering curriculum has influenced broader efforts to transform computer science education at Goethe University, particularly through the development of the Center for Critical Computational Studies and the transformation of the bachelor's and master's computer science curricula.

Strategic Profile: Computation – Cognition – Complexity (CCC)

The Institute for Computer Science (IFI) has oriented its research, teaching, and knowledge transfer around an integrative profile structured by three complementary focus areas:

- Computation - Development and analysis of modern hardware and software platforms
- Cognition - Interdisciplinary understanding of comprehension in the broadest sense
- Complexity - Computational investigation of complex systems from both algorithmic and application perspectives

The systems engineering curriculum aligns directly with this strategic profile, particularly in its integration of cognitive architectures with complex computational systems.

The Dual Nature of Computational Thinking

A key philosophical foundation of the curriculum transformation is the recognition of computation's dual nature:

- As a formal, universal language rooted in logic and mathematical reasoning
- As applied to human and social systems, requiring careful consideration of context and impacts

This balanced perspective allows for the cultivation of both rigorous technical skills and critical reflection—a foundational principle of the systems engineering approach.

Sequential Integration Approach

The curriculum transformation adopts a sequential integration approach where:

- Students first master fundamental computational concepts through clear, precise teaching of formal structures
- Abstract thinking and formal reasoning skills are developed with appropriate cognitive space
- Critical perspectives are introduced at appropriate stages as students gain technical competence
- Students learn to connect formal computational methods with real-world contexts

This approach respects both the universal, formal nature of computational thinking and the importance of ethical, social, and contextual considerations in its application.

Preparing for Digital Twins and Computational Modeling

Following recommendations from the National Academy of Engineering, the curriculum prepares students for the interdisciplinary challenges of digital twins and computational modeling:

- Integrating systems engineering, systems thinking, and system architecting
- Developing skills in data analytics, machine learning, and AI within real-world contexts

- Teaching statistical and probabilistic modeling with uncertainty quantification
- Creating interdisciplinary project opportunities that span traditional boundaries

This focus on digital twins connects all three CCC perspectives: it requires scalable computing platforms (Computation), enables understanding of dynamic complex systems (Cognition), and provides insights into systemic relationships (Complexity).

Future Directions: Safe Human-AI Co-Evolution

Looking to the future, our curriculum continues to evolve in response to emerging challenges and opportunities in AI systems engineering. Central to this evolution is the question of how to ensure safe co-evolution of humans and AI technologies.

GenAI and Future Education

The emergence of generative AI presents both challenges and opportunities for systems engineering education:

- Integrating LLMs as tools for learning and creation
- Developing frameworks for evaluating and governing genAI applications
- Exploring the limits of current LLMs in reasoning and understanding
- Training students to design systems that complement human intelligence

Our ALI project has begun exploring these questions through specific case studies in educational technology, but much work remains to fully understand the implications of genAI for systems engineering.

From Systems Engineering to Life-Long Learning

As AI continues to evolve rapidly, systems engineers must engage in continuous learning throughout their careers. Our curriculum is evolving to support this through:

- Focus on invariant skills and principles that transcend specific technologies
- Development of meta-learning capabilities that enable adaptation to new paradigms

- Creation of lifelong learning communities among alumni and industry partners
- Integration of current research into educational materials

Beyond Technical Systems: Philosophical Questions

The ultimate goal of our curriculum is not just to train students in technical skills but to prepare them to engage with deeper questions about the relationship between humans, technology, and intelligence:

- What forms of intelligence are possible, and which are desirable?
- How can we design systems that enhance rather than diminish human capabilities?
- What governance structures are needed for increasingly autonomous systems?
- How might AI help us better understand ourselves and our place in the world?

By fostering this philosophical dimension alongside technical mastery, we aim to prepare students not just for jobs but for lives of meaning, purpose, and responsible technological citizenship.

Conclusion: Preparing Holistic Systems Thinkers for the AI Era

As artificial intelligence transforms society, the need for engineers who understand both technical systems and their broader contexts becomes increasingly critical. The systems engineering curriculum at Goethe University Frankfurt represents a forward-thinking approach to this challenge, preparing students to navigate the complex landscape of modern AI development. Central to this approach is the recognition that effective AI systems engineering requires more than just technical expertise—it demands a holistic perspective that integrates multiple viewpoints, balances competing concerns, and considers the full lifecycle of system development and deployment. Our "(C, T, P) ---t (HW/SW Sys)" framework provides a systematic methodology for this integration, while our project-based learning approach gives students practical experience in applying these principles. The demonstrated impact of our curriculum—from academic achievements to industry placements to societal contributions—validates the effectiveness of this approach. By training "holistic systems thinkers" who combine technical depth with systemic breadth, we are addressing Germany's MINT professional shortage while contributing to the responsible development of AI technologies. As we look to the future, the question of safe human-AI co-evolution becomes increasingly central. Our curriculum continues to evolve in response to this challenge, exploring how systems engineering principles can help design technologies that enhance rather than diminish human capabilities. By combining technical rigor with philosophical depth, we aim to prepare students not just for careers but for leadership in shaping a future where technology serves human flourishing. In the words of one of our guiding principles, systems thinking is central to safe AI design. Through our educational approach, we are cultivating systems thinking in the next generation of engineers and computer scientists, equipping them to create AI systems that are not only powerful but also transparent, fair, and aligned with human values.

Mitochondria Under Pressure: Exploring the Stress Mind-Body Connection in Health and Disease

Dr. Darshan Mehta

Assistant Professor, Harvard Medical School, Boston, USA

Abstract

Emerging research reveals mitochondria as central mediators in the bidirectional relationship between psychological stress and physiological health, offering new insights into mind-body medicine. Mitochondria dynamically regulate stress responses through energy production, redox signaling, and interorgan communication. Chronic stress induces *mitochondrial allostatic load* — structural and functional recalibrations that impair energy metabolism, alter neuroendocrine signaling (HPA axis and SAM activation), and promote systemic inflammation. These mitochondrial changes create distinct multisystemic stress signatures that influence disease susceptibility. Mind-body interventions appear to modulate these pathways through 1) **Stress hormone regulation**, reducing cortisol and catecholamine levels that drive mitochondrial dysfunction; 2) **Cellular repair mechanisms**, enhancing mitochondrial biogenesis and dynamics via nitric oxide signaling; 3) **Metabolic optimization**, improving nutrient sensing and redox balance through dietary interventions; and 4) **Neural plasticity**, promoting hippocampal mitochondrial adaptations through meditation and relaxation practices. Experimental models show that mitochondrial genotype influences stress reactivity patterns, while clinical studies suggest mind-body practices may mitigate mitochondrial allostatic load by enhancing stress resilience. This synthesis positions mitochondrial health as both a biomarker and therapeutic target in personalized approaches to stress-related disorders.

Introduction

Psychological stress is a major risk factor for a wide spectrum of chronic conditions, including cardiovascular disease, depression, metabolic syndrome, and neurodegeneration (Picard & McEwen, 2018a). Classical stress models have emphasized neuroendocrine mediators such as the hypothalamic–pituitary–adrenal (HPA) axis and sympathetic nervous system, but they do not fully explain how subjective experiences become biologically embedded across tissues and over time (Kelly et al., 2012; Picard & McEwen, 2018a).

Mitochondria—multifunctional organelles responsible for ATP production, redox signaling, steroidogenesis, and apoptosis—have emerged as a crucial interface between psychosocial experiences and systemic physiology (Picard & McEwen, 2018a, 2018b). Conceptual and empirical work suggests that mitochondria both respond to and shape psychological states, enabling a bidirectional mind–mitochondria axis that participates in stress regulation, resilience, and disease vulnerability (Picard & McEwen, 2018; Trumpff et al., 2024).

The concept of mitochondrial allostatic load provides a unifying framework to understand how chronic stress induces cumulative mitochondrial "wear and tear" through sustained neuroendocrine and immune signaling, with downstream endocrine, metabolic, and neurocognitive consequences (Picard et al., 2014; Kelly et al., 2012). At the same time, converging data indicate that mind–body practices and targeted lifestyle interventions can recalibrate mitochondrial function and reduce MAL, creating the basis for personalized mitochondrial medicine (Gautam et al., 2021; Khurana et al., 2021; Li et al., 2023).

The Mind–Mitochondria Axis

Mitochondria as Stress Sensors and Transducers

Preclinical and human data indicate that mitochondria are dynamic sensors of psychosocial stress that integrate neuroendocrine signals into changes in energy production, redox status, and inflammatory signaling (Picard & McEwen, 2018a, 2018b). In animal models, acute and chronic psychological stress paradigms alter mitochondrial morphology, respiratory chain activity, and oxidative phosphorylation capacity in brain and peripheral tissues (Picard & McEwen, 2018b). In humans, chronic psychosocial stress and psychopathology have been associated with alterations in mitochondrial DNA copy number, respiratory chain enzyme activity, and cellular respiration in blood cells, suggesting systemic mitochondrial recalibration (Picard & McEwen, 2018a).

A 2024 study using human post-mortem brain tissue reported that positive psychosocial experiences across the lifespan correlated with a higher abundance of mitochondrial oxidative phosphorylation machinery in specific cortical regions. In contrast, negative experiences were associated with reduced mitochondrial energy capacity and altered components of oxidative phosphorylation (Trumpff et al., 2024). These findings support the view that psychosocial environments exert durable, region-specific effects on brain mitochondrial biology.

Bidirectional Feedback Between Mitochondria and Mental States

Conceptual frameworks and empirical data support a bidirectional relationship in which psychological stress alters mitochondrial function, and mitochondrial status in turn influences mood, cognition, and stress reactivity (Picard & McEwen, 2018a, 2018b). Human observational work shows that day-to-day fluctuations in mood predict subsequent changes in composite indices of mitochondrial health derived from blood-based measures, implying that emotional states dynamically influence mitochondrial function (Picard & McEwen, 2018a).

Conversely, mitochondrial dysfunction—through energy deficits, increased reactive oxygen species, and altered redox signaling—can modulate neurotransmission, synaptic plasticity, and neuroinflammation, thereby contributing to depressive symptoms and cognitive impairment (Venkatesan et al., 2026). This reciprocal feedback helps explain why chronic stress can become self-perpetuating, as mitochondrial impairment undermines the neural resources needed for effective stress regulation.

Mitochondrial Allostatic Load: Concept and Stages

Defining Mitochondrial Allostatic Load

Allostatic load refers to the cumulative physiological burden imposed by repeated or chronic activation of allostatic systems in response to stress (Kelly et al., 2012). Extending this framework, mitochondrial allostatic load is defined as the ensemble of molecular, structural, and functional changes that mitochondria undergo in response to chronic stress exposures, including alterations in ATP synthesis, reactive oxygen species production, fusion–fission dynamics, substrate preference, and signaling molecule release (Picard et al., 2014; Picard & McEwen, 2018b).

Unlike traditional allostatic biomarkers, which are individual molecules (e.g., cytokines), mitochondria are complex organelles whose stress-induced recalibrations can exhibit non-linear, often inverted-U-shaped responses—adaptive at moderate levels but detrimental when chronically exaggerated or insufficient (Picard et al., 2014). MAL thus captures both quantitative shifts in mitochondrial parameters and qualitative reprogramming of mitochondrial behavior that collectively influence cellular resilience or vulnerability (Picard et al., 2014; Kelly et al., 2012).

Three-Stage Model of Stress-Induced Mitochondrial Damage

Conceptual and experimental work suggests a three-stage trajectory of stress-related mitochondrial alterations, mapping onto the broader allostatic framework (Picard et al., 2014; Picard & McEwen, 2018b).
First, during an acute adaptive response, short-term stressors increase energy demand and trigger rapid mitochondrial shape changes and functional upregulation, including transient increases in membrane potential, calcium-buffering capacity, and resistance to apoptosis (Picard & McEwen, 2018b). Glucocorticoids acutely enhance mitochondrial energy production and may promote biogenesis, expanding the dynamic range of stress responses (Picard et al., 2014). Second, with sustained or repetitive stress, persistent reactive oxygen species generation combined with overwhelmed repair systems leads to oxidative damage to mitochondrial DNA, proteins, and lipids, as well as dysregulated fusion–fission dynamics and compromised quality control (Picard & McEwen, 2018b; Venkatesan et al., 2026). Third, damaged mitochondria release danger-associated molecular patterns, including cell-free mitochondrial DNA, which activate innate immune receptors and NF-κB–dependent cytokine production, thereby propagating low-grade systemic inflammation and contributing to chronic disease phenotypes (Kelly et al., 2012; Picard & McEwen, 2018b; Venkatesan et al., 2026).

Neuroendocrine–Mitochondrial Crosstalk

Psychological stress activates two major neuroendocrine systems: the sympathetic nervous system and the HPA axis, leading to the release of circulating catecholamines and glucocorticoids (Picard & McEwen, 2018a). These hormones act not only at receptors in target tissues but also directly on mitochondria, which express glucocorticoid and other steroid receptors and are sensitive to catecholamine-regulated calcium fluxes (Picard et al., 2014; Picard & McEwen, 2018b).

Glucocorticoids acutely enhance mitochondrial membrane potential, calcium buffering, and oxidative phosphorylation, but chronic exposure can lead to mitochondrial DNA damage, impaired biogenesis, and a decline in ATP-generating capacity (Picard et al., 2014; Picard & McEwen, 2018b). Catecholamine-induced disturbances in mitochondrial calcium handling can acutely modulate respiratory chain activity and, when persistent, promote opening of the permeability transition pore, reactive oxygen species overproduction, and apoptotic signaling (Venkatesan et al., 2026).

Because many endocrine pathways rely on mitochondrial ATP and steroidogenic enzymes, MAL reverberates across hormonal systems (Kelly et al., 2012; Picard & McEwen, 2018b). Adrenal steroidogenesis requires ATP-dependent enzymatic steps; mitochondrial dysfunction in steroidogenic cells can lead to blunted or dysregulated cortisol rhythms, reinforcing HPA axis instability (Picard et al., 2014). The activation of thyroxine to triiodothyronine and downstream metabolic actions of thyroid hormones depend on mitochondrial integrity, so MAL can manifest as functional hypothyroidism and reduced basal metabolic rate (Picard & McEwen, 2018a). Similarly, ATP production in pancreatic β-cells is essential for glucose-stimulated insulin secretion, and mitochondrial dysfunction impairs this coupling, contributing to dysglycemia and metabolic syndrome (Venkatesan et al., 2026).

Mitochondria, Brain Energy, and Psychological Symptoms

High-energy-demand brain regions, such as the hippocampus and prefrontal cortex, are especially vulnerable to mitochondrial dysfunction and stress-related hypometabolism (Picard & McEwen, 2018b; Trumpff et al., 2024). In animal models, chronic stress reduces mitochondrial respiratory capacity and alters morphology in hippocampal and cortical neurons, which correlates with impairments in learning, memory, and affect regulation (Picard & McEwen, 2018b). Human neuroimaging and post-mortem studies in depression and other stress-related disorders reveal regional reductions in metabolic rate, changes in the abundance of mitochondrial oxidative phosphorylation proteins, and altered mitochondrial DNA content, consistent with a state of mitochondrial hypofunction (Trumpff et al., 2024; Venkatesan et al., 2026).

These bioenergetic constraints manifest clinically as cognitive fatigue, slowed processing speed, and executive dysfunction, which further undermine coping capacity and resilience (Venkatesan et al., 2026). Neuroplasticity—the brain's ability to form and reorganize synaptic connections—depends on sufficient mitochondrial ATP and redox signaling to support synaptic remodeling and gene expression (Picard & McEwen, 2018b). Chronic stress reduces expression of brain-derived neurotrophic factor (BDNF) and impairs mitochondrial function in neurogenic niches, thereby constraining structural and functional plasticity (Venkatesan et al., 2026).

Mind–Body and Lifestyle Interventions Targeting Mitochondria

Mind–body and lifestyle interventions can beneficially modulate several of the pathways that contribute to MAL. Randomized trials and meta-analyses indicate that meditation, yoga, tai chi, and related practices reduce basal cortisol, improve diurnal cortisol rhythmicity, and decrease sympathetic overactivity, as evidenced by reductions in resting heart rate and blood pressure (Gautam et al., 2021; Khurana et al., 2021). By dampening chronic activation of the HPA axis and sympathetic nervous system, these interventions lessen the neuroendocrine drive that contributes to MAL (Picard & McEwen, 2018b).

Yoga and meditation have been shown to increase expression of genes involved in mitochondrial integrity and biogenesis (e.g., PGC-1α, sirtuins) and to elevate mitochondrial copy number in peripheral cells (Gautam et al., 2021; Khurana et al., 2021). For example, a clinical study of rheumatoid arthritis demonstrated that an 8-week yoga intervention improved mitochondrial health indices and reduced disease severity, consistent with enhanced bioenergetic capacity and reduced inflammatory signaling (Khurana et al., 2021).

Intermittent fasting and caloric restriction represent controlled metabolic stressors that activate AMPK–sirtuin–PGC-1α signaling, promoting mitochondrial biogenesis, improved respiratory efficiency, and enhanced autophagy and mitophagy (Li et al., 2023; Zhang et al., 2025). In adipose and other tissues, intermittent fasting has been shown to increase sirtuin expression, shift fusion–fission balance toward fusion, and augment ATP production (Li et al., 2023). Sirtuins more broadly act as key gatekeepers of mitophagy and mitochondrial quality control, linking cellular energy status to the selective removal of damaged mitochondria (Lagunas-Rangel, 2025).

Exercise-based interventions also produce robust mitochondrial adaptations. A recent synthesis of work on mitochondrial function and depression highlights that regular physical activity promotes specific mitochondrial changes—via BDNF, AMPK, and PGC-1α signaling—that counteract oxidative stress, systemic inflammation, and impaired neuroplasticity, thereby reducing depressive symptoms (Silva et al., 2025; Zhang et al., 2025).

Toward Personalized Mitochondrial Medicine

Inter-individual variability in mitochondrial DNA haplogroups and in nuclear genes encoding respiratory chain components influences mitochondrial efficiency, reactive oxygen species production, and vulnerability to stress-induced damage (Picard & McEwen, 2018b; Venkatesan et al., 2026). Certain mitochondrial polymorphisms have been associated with a greater risk of stress-related disorders and may modulate responsiveness to interventions that target mitochondrial pathways (Venkatesan et al., 2026). As genomic profiling becomes more accessible, mitochondrial and nuclear-encoded mitochondrial gene variants could inform stratification of patients into subgroups with distinct mitochondrial phenotypes, such as high reactive oxygen species producers or those with impaired biogenesis capacity.

Advances in mitochondrial biomarkers support the development of MAL indices that integrate multiple functional and molecular parameters, including mitochondrial DNA copy number, respiratory chain enzyme activities, membrane potential, and levels of cell-free mitochondrial DNA (Kelly et al., 2012; Picard & McEwen, 2018a). Prospective studies incorporating standardized mitochondrial biomarkers into clinical trials of mind–body and lifestyle interventions are beginning to clarify how changes in MAL relate to symptom trajectories and resilience, for example, in conditions such as treatment-resistant depression (Silva et al., 2025).

In this context, personalized mitochondrial medicine for stress-related disorders can be conceptualized as a multistep process. First, phenotype assessment involves comprehensive clinical characterization of stress-related symptomatology, metabolic profile, cardiovascular status, sleep patterns, and cognitive–emotional functioning to map potential mitochondrial vulnerabilities (Picard & McEwen, 2018a). Second, biomarker profiling uses mitochondrial and related biomarkers to quantify MAL and identify dominant pathophysiological pathways, such as redox imbalance versus biogenesis failure (Kelly et al., 2012; Venkatesan et al., 2026). Third, intervention matching combines mind–body practices (for example, meditation, yoga, breathing techniques), structured exercise, nutritional strategies (for example, time-restricted eating, anti-inflammatory diets), and targeted mitochondrial nutrients or pharmacologic agents (such as NAD precursors, mitophagy modulators, and specific antioxidants) to address the predominant MAL drivers in a given individual (Gautam et al., 2021; Li et al., 2023; Lagunas-Rangel, 2025; Zhang et al., 2025). Fourth, longitudinal monitoring of clinical outcomes and mitochondrial biomarkers enables dynamic adjustment of the intervention plan, aiming to progressively reduce MAL, stabilize endocrine and inflammatory markers, and enhance resilience and quality of life (Picard & McEwen, 2018a).

Conclusion

Mitochondria occupy a central position in the translation of psychosocial stress into cellular and systemic pathology, functioning as both targets and mediators of neuroendocrine and immune processes. The construct of mitochondrial allostatic load captures the cumulative molecular, structural, and functional alterations that underlie stress-related bioenergetic failure, endocrine disruption, neurocognitive symptoms, and chronic inflammation. Converging evidence from basic science, clinical studies, and emerging multi-omic analyses demonstrates that mind–body and lifestyle interventions can beneficially modulate mitochondrial function by normalizing stress hormone signaling, enhancing biogenesis and mitophagy, restoring redox balance, and promoting neuroplasticity.

These findings support a shift from symptom-focused stress management toward integrative, mitochondria-centered approaches that address core cellular mechanisms of disease and resilience. As mitochondrial biomarkers, genetic profiling, and mechanistically informed interventions continue to evolve, personalized mitochondrial medicine offers a promising pathway to tailor mind–body and metabolic strategies to individual stress phenotypes, ultimately improving outcomes across a range of stress-related conditions.

References

Gautam, S., Saxena, R., Dada, T., & Dada, R. (2021). Yoga—impact on mitochondrial health: Clinical consequences. Annals of Neurosciences, 28(3–4), 114–116.

Kelly, M., Tyrka, A. R., Price, L. H., & Carpenter, L. L. (2012). Mitochondrial nexus to allostatic load biomarkers. Psychosomatic Medicine, 74(9), 799–809.

Khurana, R., Choudhary, R., KJ, M., Dada, T., & Dada, R. (2021). Yoga improves mitochondrial health and reduces severity of rheumatoid arthritis: A randomized controlled trial. Complementary Therapies in Clinical Practice, 43, 101312.

Lagunas-Rangel, F. A. (2025). Sirtuins in mitophagy: Key gatekeepers of mitochondrial quality. Molecular and Cellular Biochemistry, 480(1–2), 67–84.

Li, X., Gao, R., Chen, W., et al. (2023). Intermittent fasting promotes adipocyte mitochondrial fusion through Sirt3 deacetylation of Mdh2. British Journal of Nutrition, 130(7), 1223–1236.

Picard, M., Juster, R.-P., & McEwen, B. S. (2014). Mitochondrial allostatic load puts the "gluc" back in glucocorticoids. Nature Reviews Endocrinology, 10(5), 303–310.

Picard, M., & McEwen, B. S. (2018a). Psychological stress and mitochondria: A conceptual framework. Psychosomatic Medicine, 80(2), 126–140.

Picard, M., & McEwen, B. S. (2018b). Psychological stress and mitochondria: A systematic review. Psychosomatic Medicine, 80(2), 141–153.

Silva, T. F., et al. (2025). Adaptations in mitochondrial function induced by exercise. Journal of Affective Disorders, 345, 215–228.

Trumpff, C., et al. (2024). Psychosocial experiences are associated with the human brain's mitochondrial oxidative phosphorylation. Proceedings of the National Academy of Sciences of the United States of America, 121(25), e2317673121.

Venkatesan, S., et al. (2026). Mitochondrial dysfunction: The cellular bridge from emotional stress to disease onset. Biomolecules, 16(1), 117.

Zhang, Y., et al. (2025). AMPK/SIRT1/PGC-1α signaling pathway: Molecular mechanisms and targeted strategies from energy homeostasis regulation to disease therapy. CNS Neuroscience & Therapeutics, 31(11), e70657.

Bringing the Body Back into Mind-Body (and Consciousness) Research

Dr. Peter M. Wayne

Osher Center for Integrative Health, Harvard Medical School, and Brigham and Women's Hospital, Boston, MA, United States

Abstract

Contemporary integrative health rejects a strict Cartesian separation between mind and body, instead emphasizing dynamic, bidirectional interactions between brain, body, and environment. Within this framework, movement, posture, and visceral signaling are understood not merely as outputs of neural processes but as active contributors to cognition, emotion, and health. Yet much research on mind–body practices continue to prioritize "top-down" neural mechanisms while underutilizing the rich information embedded in how bodies move and balance in real-world contexts. This paper synthesizes converging evidence that multimodal mind–body practices such as Tai Chi and Qigong function as systems-level interventions that reshape mechanical, physiological, cognitive, and affective processes, and that those embodied measures—especially gait, posture, and whole-body kinematics—are powerful, underused markers of whole-person health.

Randomized trials and meta-analyses demonstrate that Tai Chi reduces falls and fear of falling in older adults, favorably alters dual-task gait variability and cognitive function, and improves functional capacity and quality of life in patients with chronic pain and cardiorespiratory conditions. Gait speed and related parameters, in turn, predict incident dementia and capture risk within motoric cognitive risk syndromes, while postural and gait signatures reflect and shape emotional states such as depression, fear, and sickness. Emerging work in breast cancer survivors and other populations suggests that Qigong-based interventions can reduce pain and distress while enhancing interoceptive awareness and body trust, partly through changes in body shape and movement. Finally, studies manipulating posture during neurophysiologic recording show that body position itself alters cortical activity, underscoring the need to interpret brain measures in light of embodied context. Taken together, these data support a "body-into-mindbody" paradigm in which mind–body therapies are conceptualized and measured as embodied, ecological practices rather than disembodied brain training.

Keywords: Tai Chi, Qigong, gait, posture, interoception, integrative health, embodiment

Introduction

Classical Cartesian dualism, epitomized by the dictum "I think, therefore I am," has historically privileged the brain as the seat of mind and agency, relegating the rest of the body to a passive substrate. Integrative health frameworks instead emphasize systems and ecological views of health, in which psychological, physiological, and social processes are deeply interwoven and mutually constraining (Osypiuk et al., 2018; Schmalzl et al, 2014). Within this perspective, not only do cognitive and emotional states influence bodily physiology, but movement patterns, visceral signaling, and body shape continuously modulate perception, affect, and cognition (Michalak et al., 2009; Payne and Crane-Godreau, 2013).

Despite this conceptual shift, a subtle hierarchy that places brain activity at the apex of explanation remains pervasive, even within mind–body research. Conference programs and empirical reports often focus on neural correlates of meditation and contemplative practice, with relatively little attention paid to the "neck-down" body as a source of mechanistic insight and clinically relevant outcomes. This top-down bias has practical consequences: it narrows the range of metrics used to evaluate mind–body interventions, underestimates the intelligence of the body as a distributed information-processing system, and risks mischaracterizing the targets of traditional contemplative and movement-based practices (Osypiuk et al., 2018; Schmalzl et al 2014).

The present paper, inspired by recent work advocating for a more embodied approach to mind–body research, aims to articulate and support a greater appreciation of the role of the body in contemporary mind-body research. Specifically, it (a) frames Tai Chi and Qigong as prototypical multimodal, systems-level mind–body interventions; (b) reviews evidence that gait and posture are integrative markers of cognitive, affective, and physical health; (c) summarizes data on cognitive–motor coupling and dual-task paradigms in relation to mind–body training; (d) examines bidirectional links between emotion and whole-body movement; and (e) highlights emerging work on sickness behavior and posture-dependent neurophysiology that further underscores the centrality of embodied measures.

Multimodal Mind–Body Practices as Systems Interventions

Tai Chi and Qigong are traditional East Asian mind–body practices that integrate slow, intentional movements with breath regulation, focused attention, imagery, and often explicit philosophical instruction, typically delivered in group settings (Wayne and Kaptchuk, 2004a, b; Wayne et al., 2013). Conceptually, they target multiple domains simultaneously: aerobic conditioning, musculoskeletal strength and flexibility, sensorimotor integration, interoception, attentional control, affect regulation, and social connectedness. This multimodality contrasts sharply with the single–active-ingredient model of pharmacologic interventions, in which a molecule targets a specific receptor and downstream pathway.

Multimodal effects are particularly evident in fall prevention. In community-dwelling older adults, three sessions per week of Tai Chi over six months reduced the risk of multiple falls by approximately 50% relative to stretching controls and improved a range of balance and physical performance measures (Li et al., 2018). Purported mechanisms include Tai Chi-mediated improvements in lower extremity strength and flexibility, proprioception and motor coordination, executive function, and dual task performance. More recent systematic reviews and meta-analyses confirm that Tai Chi reduces both the proportion of older adults who fall at least once and the overall rate of falls, with risk ratios around 0.70–0.80 and incidence rate ratios near 0.69 (Zou et al., 2023).

Beyond physical performance, Tai Chi exerts robust effects on fear of falling and balance confidence. Meta-analyses and randomized controlled trials indicate that Tai Chi can reduce fear of falling by roughly 30–50% and improve dynamic balance, while concurrently lowering fall incidence in older adults (Liu et al., 2017; Liang et al., 2023). Together, these data illustrate that Tai Chi and related practices modulate multiple interacting risk factors for falls—strength, flexibility, sensory function, executive function, and affective responses to instability—consistent with a systems intervention rather than a single-domain exercise (Wayne et al., 2013).

Gait and Postural Control as Integrative Markers of Health

Gait and posture represent the convergent output of neuromuscular, sensory, cardiovascular, and cognitive systems interacting with environmental constraints, making them natural candidates for integrative health markers. Gait speed, in particular, has emerged as a "vital sign" of aging, predicting all-cause mortality, functional decline, and incident cardiovascular events in multiple cohorts, and more recently dementia risk (Verghese et al., 2019; Loprinzi et al., 2022). In a large prospective analysis, combined measures of gait speed and grip strength were effective, low-cost predictors of future dementia, with declines in gait speed conferring greater risk than declines in grip strength and the combination of both providing the strongest prediction (Loprinzi et al., 2022).

Motoric cognitive risk (MCR) syndrome operationalizes the coupling of mobility and cognition by defining a predementia state characterized by subjective cognitive complaints and slow gait in the absence of overt dementia (Verghese et al., 2019). Longitudinal data from pooled cohorts indicate that individuals with MCR have a substantially elevated risk of developing dementia compared to peers without MCR, and that very slow gait identifies a subgroup with particularly high risk (Verghese et al., 2019). Although not all studies find that variation in gait speed within the already slow MCR range independently predicts conversion, the syndrome as a whole appears to have greater predictive validity for dementia than slow gait alone, suggesting that combined mobility and cognitive markers capture underlying brain pathologies more effectively than either domain in isolation (Verghese et al., 2019; Loprinzi et al., 2022).

Beyond average gait speed, temporal and spatial gait parameters such as stride time variability, rhythmicity, and regularity provide complementary information about central and peripheral nervous system integrity. Increased stride time variability, for example, is associated with Parkinson's disease, multiple sclerosis, depression, and fear of falling, and predicts future falls and mobility disability in older adults (Hausdorff et al., 2004; Brandler et al., 2012). Importantly, these measures are sensitive to cognitive load, making them attractive readouts of embodied cognitive health in dual-task paradigms (Wayne et al., 2015).

Cognitive–Motor Interactions, Dual-Task Gait, and Mind–Body Training

The interdependence of cognition and motor control is well established in aging and neurology research, where executive dysfunction is associated with impaired balance, increased falls, and abnormal gait under cognitive load (Hausdorff et al., 2004; Brandler et al., 2012). Dual-task paradigms—in which individuals walk while performing a concurrent cognitive task such as serial subtraction—reveal "dual-task costs" in the form of slowed gait, increased stride time variability, and destabilized balance, even when single-task walking remains intact (Wayne et al., 2015). The magnitude of these dual-task costs is linked to fall risk and may signal early cognitive vulnerability before overt dementia is clinically apparent (Mirelman et al 2014, Mirelman et al 2017, Montero-Odessa et al 2017).

Tai Chi training appears to modify these cognitive–motor interactions. In a hybrid cross-sectional and longitudinal study, older Tai Chi experts with an average of 25 years of practice demonstrated lower stride time variability during dual-task walking compared with age- and sex-matched Tai Chi-naïve controls, despite similar gait speed and dual-task cognitive performance (Wayne et al., 2015). When Tai Chi-naïve participants were subsequently randomized to six months of Tai Chi versus usual care, those assigned to Tai Chi showed within-group reductions in dual-task stride time variability. In contrast, controls showed minimal change (Wayne et al., 2015). These findings support the hypothesis that mind–body movement training enhances the "intelligence" of movement—improving the stability and regularity of gait under cognitive challenge, potentially via strengthened executive control and sensorimotor integration.

Complementary evidence indicates that Tai Chi can directly improve cognitive performance. Preliminary studies in older adults have reported improvements in executive function measures after 10 weeks of Tai Chi training (Penedo et al., 2008; Wayne et al 2014; Wang et al 2025). More recent randomized controlled trials and quasi-experimental studies, including work in middle-aged and older adults, suggest that 24-form simplified Tai Chi and longer-term Tai Chi programs enhance executive inhibitory control, memory, and global cognition, in some cases accompanied by increased frontal electroencephalographic activation or changes in serum brain-derived neurotrophic factor (Xu et al., 2024; Xing et al., 2025). Collectively, these data support bidirectional influences in which cognition shapes movement and movement training, in turn, refines executive functions and their embodiment in gait and balance.

Embodied Emotion, Posture, and Gait

Emotional states manifest robustly in body posture and movement, and experimental manipulations of movement can reciprocally modulate affect. Depressed individuals, for example, characteristically exhibit slower gait, reduced arm swing, decreased vertical head motion, and a more slumped posture compared with never-depressed controls, as well as increased swing time variability and other markers of gait unsteadiness (Michalak et al., 2009; Hausdorff et al., 2004; Radovanović et al., 2014). Systematic reviews of emotional body language demonstrate that observers can reliably infer emotional categories such as sadness, fear, or happiness from minimal kinematic information, underscoring the informational richness of whole-body dynamics (Michalak et al., 2009; Mikkelsen et al 2024).

Experimental work further suggests that adopting body configurations associated with particular emotions can shift subjective mood and cognitive processing in corresponding directions. Gait patterns and upper-body postures mimicking sadness can increase negative affect and bias memory toward negative material, whereas more expansive, upright postures are associated with greater confidence and less self-reported distress (Wilkes et al., 2017; Nair et al., 2015). These findings align with broader embodied cognition theories and with observations from contemplative traditions that the physical "forms" of practice—such as upright, open postures in meditation or Tai Chi—are not merely vehicles for an internal mental state but integral to the cultivation of that state (Michalak et al., 2009).

Fear of falling provides a clinically salient example of embodied emotion. Older adults with high fear of falling often adopt guarded, flexed postures, shortened step lengths, and reduced gait speed, patterns that both reflect anxiety and paradoxically increase fall risk by degrading balance and discouraging activity (Liu et al., 2017; Sloman et al., 1982). Randomized controlled trials and meta-analyses demonstrate that Tai Chi reduces fear of falling and improves balance confidence, with these affective changes partially mediating reductions in falls (Liu et al., 2017; Liang et al., 2023; Wu et al., 2017).

Interoception, Body Trust, and Qigong in Cancer Survivorship

Breast cancer survivors with persistent post-surgical pain often face a complex constellation of physical pain, fear of recurrence, altered body image, and distrust of the body, which can lock them into a prolonged state of threat and disembodiment (Osypiuk et al, 2019). Qigong-based mind–body exercise offers a gentle, multimodal approach that simultaneously addresses pain, mood, movement, and interoceptive awareness. In a mixed-methods pilot study of a 12-week Qigong Mind–Body Exercise program for breast cancer survivors with persistent post-surgical pain, participants showed statistically and clinically meaningful improvements in pain severity and interference, fatigue, anxiety, depression, perceived stress, and multiple domains of health-related quality of life (Osypiuk et al., 2019; Osypiuk et al 2020).

Importantly, Qigong was also associated with significant gains in interoceptive awareness, including increased body listening, emotional awareness, and body trust (Osypiuk et al., 2019). Qualitative interviews revealed that participants increasingly viewed the body as a source of information and safety rather than solely as a site of threat or betrayal, describing an emerging capacity to settle internally in response to stress and to inhabit the body with greater ease (Osypiuk et al., 2020). In a separate randomized controlled trial of Chan-Chuang Qigong with breathing meditation during chemotherapy, participants in the Qigong arm exhibited improved emotional functioning and enhanced interoceptive awareness subdomains such as attention regulation and self-regulation compared with controls (Hsu et al., 2023). These findings suggest that mind–body movement practices can recalibrate the felt relationship to the body in ways that are measurable through both self-report and kinematic assessment of posture.

Sickness Behavior and Embodied Communication of Health State

Sickness behavior—characterized by fatigue, reduced activity, social withdrawal, and altered movement—provides another lens on embodied mind–body processes. Experimental induction of systemic inflammation in healthy adults using intravenous lipopolysaccharide produces not only cytokine elevations and subjective malaise but also distinctive changes in gait (Lasselin et al., 2019; Tavares et al., 2020). During the inflammatory condition, compared with placebo, participants walk with shorter, slower, and wider strides, display less arm extension and knee flexion, and hold their heads in a more downward-tilted position, with these alterations correlated with interleukin-6 levels and self-reported sickness (Lasselin et al., 2019).

Analogous findings in animal models of lipopolysaccharide-induced neuroinflammation reveal decreased locomotor activity and altered gait parameters, further supporting the link between inflammatory state and movement signatures (Hernández-Martínez et al., 2024). From an evolutionary and social perspective, such sickness-related body language may serve communicative functions—signaling vulnerability or contagion risk to conspecifics and recruiting care—while also embodying energy-conserving strategies during illness (Tavares et al., 2020). For mind–body research, these data underscore that movement patterns are sensitive readouts of internal physiological states that extend beyond musculoskeletal function to encompass immune and neuroimmune processes.

Body Posture and Neural Measures: Implications for Neuroimaging

Much contemporary research on meditation and mind–body practices relies on neuroimaging modalities such as functional magnetic resonance imaging and magnetoencephalography, which typically require participants to lie supine or remain in constrained seated postures for extended periods. However, posture itself modulates cortical activity, raising important interpretive questions for studies that equate brain signals measured in a horizontal scanner with the true neural signature of a practice (Lifshitz et al., 2017; Thibault et al., 2016; Zadeh & Basso, 2017).

Electroencephalographic studies comparing upright and supine postures show that transitioning from sitting to a horizontal position reduces high-frequency beta and gamma band activity and alters frontal asymmetry patterns, with these differences persisting over at least two hours of sustained supine rest (Zadeh & Basso, 2017). More recent work using machine learning approaches similarly reports that supine, eyes-closed posture is associated with increased alpha and theta power across frontal, central, and occipital regions compared with sitting, consistent with a deeper state of relaxation and reduced sensory engagement (Zhang et al., 2025). These posture-dependent shifts in neural activity occur independently of any explicit contemplative practice, suggesting that at least part of what is observed in scanner-based studies reflects the embodied context of measurement (Lifshitz et al., 2017; Thibault et al., 2016).

Implications and Conclusion

The convergent evidence reviewed here supports a shift from brain-centric to fully embodied models of mind–body practice and health. Multimodal interventions such as Tai Chi and Qigong demonstrably influence diverse domains—cardiorespiratory fitness, balance, gait stability, cognitive function, affect regulation, interoceptive awareness, and social engagement—by training complex patterns of movement, attention, and meaning in ecologically rich settings (Wayne and Kaptchuk, 2008; Wayne et al., 2013). Gait, posture, and whole-body kinematics emerge as especially promising integrative markers, indexing not only musculoskeletal and neurologic integrity but also cognitive load, emotional state, inflammatory status, and relational stance.

Future research should prioritize longitudinal, translational designs that bridge laboratory and real-world contexts, leveraging wearable sensors to capture gait and postural metrics during everyday activities and dual-task challenges. Trials of mind–body interventions can be strengthened by routinely including embodied outcomes—such as dual-task gait variability, fear of falling, postural alignment, and interoceptive awareness—alongside conventional symptom scales and neural measures. Mechanistic studies combining mobile neuroimaging (fNIRs), physiological monitoring (e.g., heart rate variability, inflammatory markers), and kinematics will be particularly valuable for mapping how changes in practice—breath, movement, attention, social context—translate into distributed adaptations across brain, body, and behavior.

Clinically, embracing a more embodied paradigm invites clinicians and patients to view the body not as a passive recipient of top-down regulation but as an intelligent, expressive partner in the co-creation of health and consciousness. Interventions that help individuals move with greater stability, openness, and awareness—whether through Tai Chi, Qigong, yoga, or other embodied practices—may simultaneously improve functional outcomes, mitigate cognitive decline and fall risk, alleviate mood and anxiety symptoms, and restore a sense of safety and belonging within one's own flesh. To fully realize this potential, integrative medicine research must continue to bring the body back into the center of inquiry, developing and validating embodied measures that honor the complex, ecological reality of mind–body health.

Conflict of Interest

Peter Wayne is the founder and sole owner of the Tree of Life Tai Chi Center. Peter Wayne's interests were reviewed and managed by Mass General Brigham in accordance with its conflict-of-interest policies.

References

Brandler, T. C., Wang, C., Oh-Park, M., et al. (2012). Depressive symptoms and gait dysfunction in the elderly. American Journal of Geriatric Psychiatry, 20(5), 425–432.

Hausdorff, J. M., Peng, C. K., et al. (2004). Gait unsteadiness and fall risk in depression: A preliminary study. Journal of Neurology, 251(9), 1184–1190.

Hernández-Martínez, A. M., Román, P., et al. (2024). Locomotor and gait changes in the LPS model of neuroinflammation in mice. Brain, Behavior, and Immunity, 119, 349–363.

Hsu, C. F., Lin, Y. C., Chien, C. H., & Yeh, M. L. (2023). Chan-Chuang qigong with breathing meditation improves quality of life and interoceptive awareness in patients with breast cancer: A randomized controlled trial. European Journal of Oncology Nursing, 62, 102279.

Lasselin J, Sundelin T., Wayne P. M., et al. (2019). Biological motion during inflammation in humans. Brain, Behavior, and Immunity.

Li F, Harmer P, Fitzgerald K, Eckstrom E, et al. (2018). Effectiveness of a Therapeutic Tai Ji Quan Intervention vs a Multimodal Exercise Intervention to Prevent Falls Among Older Adults at High Risk of Falling: A Randomized Clinical Trial. JAMA Intern Med, 178(10):1301-1310.

Liang, H., Qu, S., Wang, Z., Liu, J., & Chen, J. (2023). Effectiveness of Tai Chi exercise on fear of falling and balance in older adults: A meta-analysis. Journal of Exercise Science & Fitness, 21(1), 1–9.

Lifshitz M, Thibault RT, Roth RR, Raz A. (2017). Source Localization of Brain States Associated with Canonical Neuroimaging Postures. J Cogn Neurosci. 2017 Jul;29(7):1292-1301. E-pub 2017 Feb 14. PMID: 28195522.

Liu, H., Liang, Y., Wang, Z., et al. (2017). The effects of Tai Chi on fall prevention, fear of falling, and balance in older people: A meta-analysis. Archives of Gerontology and Geriatrics, 72, 195–201.

Loprinzi, P. D., Addoh, O., & Mann, J. R. (2022). Combination of gait speed and grip strength to predict cognitive decline and dementia. Alzheimer's & Dementia: Diagnosis, Assessment & Disease Monitoring, 14(1), e12353.

Michalak, J., Troje, N. F., Fischer, J., et al. (2009). Embodiment of sadness and depression—Gait patterns associated with dysphoric mood. Journal of Behavior Therapy and Experimental Psychiatry, 40(4), 536–543.

Mikkelsen M. B., Tramm G., Michalak J., et al. (2024). Bodies in action: Do contractive and expansive postures facilitate adaptive behavior? Scand J Psychol, 65(5):966-979.

Mirelman A., Maidan I., Bernad-Elazari H., et al. (2014). Increased frontal brain activation during walking while dual tasking: an fNIRS study in healthy young adults. Journal of neuroengineering and rehabilitation, 11:85.

Mirelman A., Maidan I., Bernad-Elazari H., et al. (2017). Effects of aging on prefrontal brain activation during challenging walking conditions. Brain and cognition, 115:41-46.

Montero-Odasso M. M., Sarquis-Adamson Y., Speechley M., et al. (2017). Association of Dual-Task Gait with Incident Dementia in Mild Cognitive Impairment: Results from the Gait and Brain Study. JAMA Neurology, 74(7):857-865

Nair S., Fromgar M., Sollers J., Consedine N., & Broadbent E. (2015). Do slumped and upright postures affect stress responses? A randomized trial. Health Psychology, 34(6), 632–641.

Osypiuk K., Thompson E., Wayne P. M. (2018). Can Tai Chi and Qigong Postures Shape Our Mood? Toward an Embodied Cognition Framework for Mind-Body Research. Frontiersin Human Neuroscience, May 1; 12:174.

Osypiuk K., Ligibel J., Giobbie-Hurder A., et al. (2019). Qigong Mind-Body Exercise as a Biopsychosocial Therapy for Persistent Post-Surgical Pain in Breast Cancer: A Pilot Study. Integrative Cancer Therapies, 191-12

Osypiuk K., Kilgore K., Ligibel J., et al. (2020). Making peacewith our bodies: A qualitative analysis of breast cancer survivors' experiences with Qigong mind-body exercise. J Alt Comp Med, Sep;26(9):825-832.

Payne P., Crane-Godreau M. A. (2013). Meditative movement for depression and anxiety. Front Psychiatry, Jul 24; 4:71.

Penedo, F. J., Dahn, J. R., Shen, B. J., et al. (2008). Can Tai chi enhance cognitive vitality? A preliminary study of cognitive executive control in older adults after a Tai chi intervention. Journal of Psychosomatic Research, 64(1), 55–61.

Radovanović, S., Marić, N. P., Kostić, V. S., & Aleksić, D. (2014). Gait characteristics in patients with major depressive disorder and healthy controls: A pilot study. Psychiatry Research, 218(1–2), 257–262.

Schmalzl L., Crane-Godreau M. A., Payne P. (2014). Movement-based embodied contemplative practices: definitions and paradigms. Front Hum Neurosci, Apr 14; 8:205.

Sloman, L., Pierri, J., & Barrett, F. (1982). The effects of anxiety and depression on the gait of the elderly. Comprehensive Psychiatry, 23(4), 345–351.

Sun, X., Gao, Y., Zhang, H., Yang, H., & Zhang, J. (2017). Tai Chi exercise for patients with chronic heart failure: A meta-analysis of randomized controlled trials. American Journal of Physical Medicine & Rehabilitation, 96(10), 706–716.

Tavares, G., Gidron, Y., & Zeidan, R. (2020). Biological motion during inflammation in humans. Brain, Behavior, and Immunity, 84, 28–36.

Thibault R. T., Lifshitz M., Raz A. (2016). Body position alters human resting-state: Insights from multi-postural magnetoencephalography. Brain Imaging Behav, Sep;10(3):772-80. PMID: 26409469.

Verghese, J., Wang, C., et al. (2019). Motoric cognitive risk syndrome and predictors of transition to dementia. Alzheimer's & Dementia, 15(7), 924–932.

Wang W-T., Wang H. (2025). Effects of Tai Chi Chuan on cognitive function in adults 60 years or older with mild cognitive impairment: a systematic review and meta-analysis. Frontiers in Physiology, 16:1556622.

Wayne P. M., Walsh J. N., Taylor-Piliae R. E., et al. (2014). Effect of Tai Chi on cognitive performance in older adults:

Systematic review and meta-analysis. Journal of the American Geriatrics Society, 62(1):25-39.

Wayne P. M., Kaptchuk T. J. (2008). Challenges inherent to Tai Chi Research: Part I--Tai Chi as acomplex multi-component intervention. J Alter Compl Med, 14:95-102.

Wayne P. M., Kaptchuk T. J. (2008). Challenges inherent to the Tai Chi Research: Part II -- Defining the intervention and optimal study design. J Alter Compl Me, 14:191-7.

Wayne, P. M., Kaptchuk, T. J., Peng, C. K., et al. (2004). Effects of Tai Chi mind–body movement therapy on functional status and exercise capacity in patients with chronic heart failure: A randomized controlled trial. American Journal of Medicine, 117(8), 541–548.

Wayne P. M., Manor B., & Novak V., et al. (2013). A systems biology approach to studying Tai Chi, physiological complexity, and healthy aging: Design and rationale of a pragmatic randomized controlled trial. Contemporary Clinical Trials, 34:21-34

Wayne, P. M., Gow, B. J., Costa, M. D., et al. (2015). Tai Chi training may reduce dual-task gait variability, a potential mediator of fall risk, in healthy older adults: A randomized controlled trial. Frontiers in Human Neuroscience, 9, 332.

Wilkes, C., Kydd, R., Sagar, M., & Broadbent, E. (2017). Upright posture improves affect and fatigue in people with depressive symptoms. Journal of Behavior Therapy and Experimental Psychiatry, 54, 143–149.

Wu, C., Chen, J., Huang, J., & Chen, Y. (2017). The effect of Tai Chi exercise on the risk and fear of falling in older adults: A randomized controlled trial. International Journal of Gerontology, 11(4), 210–214.

Xu, D., Cui, C., Zhang, Q., Xiao, M., & Li, Y. (2024). Effect of 24-form simplified Tai Chi on executive inhibitory control and frontal brain electrical activity in college students. Frontiers in Psychology, 15, 1362493.

Xing, X., Zhang, J., Cao, W., & Wang, L. (2025). Evaluation of the efficacy of Tai Chi on the cognitive function in older adults: A randomized controlled trial. Frontiers in Aging Neuroscience, 16, 1365432.

Yang, G., Wang, Y., Liu, Y., & Wang, H. (2017). Tai Chi for preventing falls in older adults: A systematic review and meta-analysis. BMJ Open, 7(2), e013561.

Zadeh, S. S., & Basso, J. C. (2017). Posture used in fMRI–PET elicits reduced cortical activity and altered frontal asymmetry in EEG. Frontiers in Human Neuroscience, 11, 621.

Zhang, X., Li, J., Wang, X., & Kumar, S. (2025). Machine learning approaches to evaluate EEG correlates of relaxation between supine and sitting postures. Brain Informatics, 12(1), 24.

Zou, L., Zhang, Y., Yang, L., Loprinzi, P. D., Yeung, A. S., Kong, J., & Demurtas, J. (2023). Tai Chi for fall prevention and balance improvement in older adults: A systematic review and meta-analysis. Frontiers in Public Health, 11, 12234

About INADS Press

INADS Press is the publishing arm of the Institute of Advanced Sciences (INADS), located in Dartmouth, Massachusetts, USA. As a dedicated academic publisher, INADS Press is committed to advancing scholarly discourse and promoting interdisciplinary research by publishing high-quality academic works. Our mission is to provide a platform for innovative ideas and research that contribute to global knowledge and foster a deeper understanding of the world's complex issues.

At INADS Press, we specialize in publishing a diverse range of scholarly materials, including academic books, conference proceedings, research papers, and specialized monographs. Our publications cover a wide array of disciplines, reflecting the interdisciplinary nature of research at INADS. We are particularly focused on works that integrate science, philosophy, culture, and sustainability, emphasizing both contemporary and traditional knowledge systems.

Our editorial process is rigorous, ensuring that every publication meets the highest standards of academic excellence. From peer-reviewed research to carefully edited volumes on Indic studies and cutting-edge scientific topics, INADS Press is committed to producing works that are not only academically rigorous but also accessible to a broader audience. We collaborate with scholars, researchers, and experts from around the world to bring forward publications that contribute to both academic scholarship and public knowledge.

INADS Press also places a strong emphasis on the global dissemination of knowledge. We utilize both traditional print and digital platforms to ensure that our publications reach a wide audience, including academics, professionals, and interested readers across the globe. Our commitment to quality, accessibility, and scholarly integrity makes INADS Press a trusted name in academic publishing.

For inquiries, please contact inadspress@inads.org

Made in the USA
Monee, IL
13 September 2026

62976603R00128